Moritz M. Hettich

Role of neuronal IGF-1R signaling in Alzheimer's disease

Moritz M. Hettich

Role of neuronal IGF-1R signaling in Alzheimer's disease

Involvement in development and progression

Südwestdeutscher Verlag für Hochschulschriften

Impressum/Imprint (nur für Deutschland/only for Germany)
Bibliografische Information der Deutschen Nationalbibliothek: Die Deutsche Nationalbibliothek verzeichnet diese Publikation in der Deutschen Nationalbibliografie; detaillierte bibliografische Daten sind im Internet über http://dnb.d-nb.de abrufbar.
Alle in diesem Buch genannten Marken und Produktnamen unterliegen warenzeichen-, marken- oder patentrechtlichem Schutz bzw. sind Warenzeichen oder eingetragene Warenzeichen der jeweiligen Inhaber. Die Wiedergabe von Marken, Produktnamen, Gebrauchsnamen, Handelsnamen, Warenbezeichnungen u.s.w. in diesem Werk berechtigt auch ohne besondere Kennzeichnung nicht zu der Annahme, dass solche Namen im Sinne der Warenzeichen- und Markenschutzgesetzgebung als frei zu betrachten wären und daher von jedermann benutzt werden dürften.

Verlag: Südwestdeutscher Verlag für Hochschulschriften GmbH & Co. KG
Dudweiler Landstr. 99, 66123 Saarbrücken, Deutschland
Telefon +49 681 37 20 271-1, Telefax +49 681 37 20 271-0
Email: info@svh-verlag.de

Approved by: Cologne, University of Cologne, Diss., 2009

Herstellung in Deutschland:
Schaltungsdienst Lange o.H.G., Berlin
Books on Demand GmbH, Norderstedt
Reha GmbH, Saarbrücken
Amazon Distribution GmbH, Leipzig
ISBN: 978-3-8381-2878-8

Imprint (only for USA, GB)
Bibliographic information published by the Deutsche Nationalbibliothek: The Deutsche Nationalbibliothek lists this publication in the Deutsche Nationalbibliografie; detailed bibliographic data are available in the Internet at http://dnb.d-nb.de.
Any brand names and product names mentioned in this book are subject to trademark, brand or patent protection and are trademarks or registered trademarks of their respective holders. The use of brand names, product names, common names, trade names, product descriptions etc. even without a particular marking in this works is in no way to be construed to mean that such names may be regarded as unrestricted in respect of trademark and brand protection legislation and could thus be used by anyone.

Publisher: Südwestdeutscher Verlag für Hochschulschriften GmbH & Co. KG
Dudweiler Landstr. 99, 66123 Saarbrücken, Germany
Phone +49 681 37 20 271-1, Fax +49 681 37 20 271-0
Email: info@svh-verlag.de

Printed in the U.S.A.
Printed in the U.K. by (see last page)
ISBN: 978-3-8381-2878-8

Copyright © 2011 by the author and Südwestdeutscher Verlag für Hochschulschriften GmbH & Co. KG and licensors
All rights reserved. Saarbrücken 2011

Für Dich, Mama.
Ich hoffe, dass ich Dir einen Deiner Träume für uns erfüllen konnte.
Beijo Mo

1 INTRODUCTION 9

1.1 ALZHEIMER'S DISEASE 10
1.2 HERITABLE CAUSES OF ALZHEIMER'S DISEASE 11
1.3 NEURITIC PLAQUES AND NEUROFIBRILLARY TANGLES 12
1.4 PROCESSING OF APP 13
1.5 CLEARANCE OF AMYLOID BETA (Aβ) 15
1.6 SECRETASES 16
1.7 IGF-1 SIGNALING CASCADE 18
1.7.1 INSULIN AND INSULIN-LIKE-GROWTH FACTOR-1 SIGNALING IN ALZHEIMER'S DISEASE 18
1.7.2 IGF-1 AND IGF-1R 18
1.7.3 IGF-1R/IR SIGNALING 19
1.8 MOUSE MODELS 22
1.8.1 CONDITIONAL IGF-1R KNOCK OUT (THE CRE/LOXP SYSTEM) 22
1.8.2 CRE RECOMBINASE EXPRESSION UNDER THE CONTROL OF THE SYNAPSIN-1 PROMOTOR 22
1.8.3 THE ALZHEIMER'S DISEASE MODEL TG2576 23
1.9 AIMS OF THIS THESIS 24

2 MATERIAL AND METHODS 25

2.1 CHEMICALS 26
2.1.1 BUFFER AND SOLUTION 28
2.1.2 KITS 29
2.1.3 PRIMARY ANTIBODIES 30
2.1.4 SECONDARY ANTIBODIES 32
2.2 MATERIALS 33
2.3 METHODS 34
2.3.1 ISOLATION OF GENOMIC DNA 34
2.3.2 QUANTIFICATION OF NUCLEIC ACID 34
2.3.3 POLYMERASE CHAIN REACTION (PCR) 34
2.3.4 ANIMALS, BREEDING AND GENOTYPING 35
2.3.5 HISTOLOGY AND IMMUNOSTAINING 36
2.3.6 METABOLIC CHARACTERIZATION, GLUCOSE, AND INSULIN TOLERANCE TESTS 36
2.3.7 ANALYSIS OF BODY COMPOSITION 37
2.3.8 ISOLATION OF CEREBELLAR GRANULE CELLS 37
2.3.9 IMMUNOBLOTTING 38
2.3.10 GEL ELECTROPHORESIS 38
2.3.11 WESTERN BLOT 39
2.3.12 SECRETASE ACTVITY ASSAYS 42
2.3.13 ELISA β-AMYLOID$_{1-40/42}$ 42
2.3.14 STATISTICAL ANALYSIS 43

3 RESULTS 44

3.1 IGF1R EXPRESSION IN CEREBELLAR GRANULE CELLS OF NEURON-SPECIFIC IGF-1R KNOCKOUT MICE (NIGF-1R$^{-/-}$) 46
3.2 PATTERN OF SYNAPSIN-1 PROMOTER DRIVEN CRE RECOMBINASE ACTIVITY IN THE CNS 47
3.3 IGF-1R EXPRESSION IN THE CNS AND PERIPHERAL TISSUES OF NIGF-1R$^{-/-}$ MICE 49

3.4 IGF-1R SIGNALING IN HIPPOCAMPUS AFTER ACUTE IGF-1 STIMULATION   51
3.5 KAPLAN-MEIER ANALYSIS   52
3.6 METABOLIC AND SOMATIC CHARACTERISATION   55
3.6.1 GLUCOSE HOMEOSTASIS   56
3.6.2 SOMATIC CHARACTERISATION   58
3.7 BIOCHEMICAL ANALYSIS OF 28 WEEKS OLD ANIMALS   64
3.7.1 ANALYSIS OF IGF-1R/IR SIGNALING   64
3.7.2 INVESTIGATION OF APP PROCESSING   66
3.8 BIOCHEMICAL ANALYSIS OF 60 WEEKS OLD ANIMALS   69
3.8.1 ANALYSIS OF IGF-1R/IR SIGNALING   69
3.8.2 INVESTIGATION OF THE APP PROCESSING   73

4 DISCUSSION   81

4.1 TG2576 MOUSE MODEL AND NEURON-SPECIFIC IGF1-R DELETION   82
4.2 METABOLIC CHARACTERISATION   83
4.3 SOMATIC CHARACTERISATION   84
4.4 SURVIVAL AND AGING   85
4.5 BIOCHEMICAL ANALYSIS OF THE IGF-1R/IR SIGNALING AND APP METABOLISM 86
4.6 PERSPECTIVES AND EXPERIMENTAL APPROACH   89

5 SUMMARY   91

6 REFERENCES   94

Figure Index

Fig. 1-1 Illustration of APP processing by α-, β- and γ-secretases 14
Fig. 1-2 Illustration of IGF-1R/IR signaling cascade 20
Fig. 3-1 Illustration of the breeding strategy 46
Fig. 3-2 Cerebellar granule cells of nIGF-1R$^{-/-}$ mice 47
Fig. 3-3 β-Galactosidase staining representing Cre recombinase activity in synCre lacZ reporter mice 48
Fig. 3-4 β-galactosidase staining representing Cre recombinase activity in the hippocampal formation of synCre lacZ reporter mice 49
Fig. 3-5 Western blot analysis of IGF-1R protein expression in different brain regions 50
Fig. 3-6 Densitometric quantification of IGF-1R expression in the CNS 51
Fig. 3-7 Western blot analysis of IGF-1R protein expression in peripheral tissues 51
Fig. 3-8 Western blot analysis of IGF-1R expression of Hippocampus and Cortex 52
Fig. 3-9 Kaplan-Meier analysis of WT, Tg2576, nIGF-1R$^{-/-}$ and nIGF-1R$^{-/-}$Tg2576 animals 53
Fig. 3-10 Kaplan-Meier analysis of WT, Tg2576, nIGF-1R$^{-/-}$ and nIGF-1R$^{-/-}$Tg2576 females and males 54
Fig. 3-11 Kaplan-Meier analysis of WT, Tg2576, nIGF-1R$^{+/-}$ and nIGF-1R$^{+/-}$Tg2576 animals 55
Fig. 3-12 Blood glucose levels of male and female mice during 60 weeks of observation 56
Fig. 3-13 Glucose tolerance test of male and female mice 57
Fig. 3-14 Insulin tolerance test from male and female mice 57
Fig. 3-15 Body length of females and males of the study group 58
Fig. 3-16 Brain weight of females and males of the study group 59
Fig. 3-17 Body weight of 60 weeks old animals and growth curves of the different genotypes 61
Fig. 3-18 Fat content at 28 weeks 61
Fig. 3-19 Fat content at 60 weeks 62
ig. 3-20 Comparison of Fat content of 28 and 60 weeks old animals 63
Fig. 3-21 Brain-Body ratio of 60 weeks old mice 63
Fig. 3-22 Western blot analysis of IGF1-R/IR signaling of 28 weeks old mice I 64
Fig. 3-23 Western blot analysis of IGF1-R/IR signaling of 28 weeks old mice II 65

Fig. 3-24 Western blot and densitometric analysis of APP processing of 28 weeks old mice 67
Fig. 3-25 Western blot and ELISA analysis of Amyloid-β in 28 weeks old mice 67
Fig. 3-26 Western blot analysis of proteins involved in APP cleavage and Aβ clearance 68
Fig. 3-27 Western blot analysis of IGF1-R/IR signaling of 60 weeks old mice 69
Fig. 3-28 Densitometric quantification of IGF-1R, IRS-1 and IRS-2 protein expression in 60 weeks old mice 70
Fig. 3-29 Western blot analysis of ERK-1/2 71
Fig. 3-30 Western blot analysis of AKT and PTEN of 60 weeks old mice 72
Fig. 3-31 Western blot analysis of GSK-3 72
Fig. 3-32 Western blot analysis of Foxo1 73
Fig. 3-33 Western blot analysis of C-teminal fragments (CTFs) 74
Fig. 3-34 Quantification of Aβ$_{1-40/42}$ in 60 weeks old Tg2576 and nIGF-1R$^{-/-}$Tg2576 75
Fig. 3-35 Histochemical staining of Amyloid plaques I 76
Fig. 3-36 Histochemical stainings of Amyloid plaques II 77
Fig. 3-37 Western blot analysis of proteins involved in clearance of Aβ 77
Fig. 3-38 Western blot analysis of α-, β- and γ-secretases in hippocampus and cortex of 60 weeks old mice 78
Fig. 3-39 α-secretase activity assay 79
Fig. 3-40 β-secretase activity assay 79

Table Index

Table 2-1 Oligonucleotides used for genotyping 35
Table 2-2 SDS-PAGE mini gels (2 x) 39

List of Abbreviations

AD	Alzheimer's disease
ADAM	A Disintegrin And Metalloprotease domain
AKT	PKB synonym
α2M	alpha 2 macroglobulin
apoE	Apolipoprotein E
APP	Amyloid Precursor Protein
APS	Ammonium-persulfate
Aβ	β-Amyloid
BACE-1	Beta-site APP Cleaving Enzyme-1
BBB	Blood brain barrier
BME	Basal medium eagle
BSA	Bovine serum albumin
C83	83-amino-acid C-terminal APP fragment
C99	99-amino-acid C-terminal APP fragment
CNS	Central Nervous System
CSF	Cerebrospinal fluid
ddH2O	Double-disalled water
DMSO	Dimethyl sulfoxide
eAD	early onset Alzheimer's disease
ELISA	Enzyme Linked Immunosorbent Assays
ER	Endoplasmic reaculum
ERK	Extracellular signal-regulated kinase
FAD	familial Alzheimer's disease
FCS	Fetal calf serum
GDP	Guanosine-diphosphate
GH	Growth hormone
GRB2	Growth factor receptor binding protein 2
GSK-3α/β	Glycogen synthase kinase 3α/β
GTP	Guanosine-triphosphate
HBSS	Hank's balanced salt solution
IDE	Insulin degrading enzyme
IGF	Insulin-like growth factor

IGF-1R	Insulin-like growth factor receptor type I
IR	Insulin Receptor
IRa	Insulin receptor isoform a
IRb	Insulin receptor isoform b
IRS-1	Insulin receptor substrate 1
IRS-2	Insulin receptor substrate 2
IRSs	Insulin receptor substrates
kDA	kilo Dalton
LOAD	Late onset of Alzheimer's disease
mA	milli Ampere
MAP-kinase	Mitogen-acavated protein kinase
MEK	Mitogen-acavated protein kinase kinase
NFTs	Neurofibrillary tangles
NIDDM	Non-insulin-dependent diabetes mellitus / type-2 diabetes
nIGF-1R$^{-/-}$	neuronal specific IGF-1R knockout
P/S	Penicillin-Streptomycin; Pen Strep
p3	Short pepade containing the C-terminal region of Aβ
PAGE	Polyacrylamide gel electrophoresis
PBS	Phosphate buffered saline
PDK1	Phosphoinosiade-dependent protein kinase 1
PDVF	Polyvinylidene difluoride
Pi3K	Phosphaadylinositol-tri-phosphat kinase
PIP2	Phosphaadylinositol-di-phosphat
PIP3	Phosphaadylinositol-tri-phosphat
PKB	Protein kinase B
PP2A	Protein phosphatase 2A
PSEN1	Presenilin 1
PSEN2	Presenilin 2
PTEN	Phosphatase and tensin homolog
rpm	Rotations per minute
sAPPα	soluble APPα
sAPPβ	soluble APPβ
SDS	Sodium dodecyl sulfate
SDS-PAGE	Sodium dodecyl sulfate-polyacrylamide gel electrophoresis

SH2	Src-homology 2
SHP2	SH2-Phosphatase 2
SOS	Son of sevenless
SPs	Senile plaques
TACE	Tumor necrosis factor-alpha converting enzyme
TBS	Tris buffered saline
TBS-T	Tris buffered saline 2% TWEEN 20®
TEMED	N,N,N',N'-tetramethylethylenediamine
Tg2576	Transgenig mouse model for Alzheimer's disease
TGN	Trans-Golgi network
TRkA	Tyrosine kinase receptor A
TWEEN 20®	Polyoxyethylene (20) sorbitan monolaurate (Polysorbate 20)

1 Introduction

1.1 Alzheimer's disease

In 1901 the german psychiatrist and neuropathologist Alois Alzheimer described a case of a middle aged woman called Auguste D. with strange behavioural symptoms and progressive loss of cognitive abilities. In 1906, after Auguste D. died, he published his first essay on this phenomenon and Auguste D. was the first person diagnosed with Alzheimer's disease (AD)[1]. AD is a chronic progressive neurodegenerative disorder resulting in death after an average of 8–10 years after diagnosis[2]. Its clinical manifestation is typified by three groups of symptoms[3].

o Cognitive dysfunction: In this group the symptoms include memory loss, language disabilities and executive dysfunction (that means, loss of higher level planning and intellectual coordination skills).

o Non-cognitive symptoms: This group of symptoms comprises psychiatric symptoms and behavioural disturbance e.g. depression, hallucinations, delusions and agitation.

o The third group includes restrictions in performing activities of everyday life (defined as "instrumental" for more complex activities such as driving and shopping and "basic" for unaided dressing and eating).

The symptoms of AD progress from minor symptoms of memory loss, mild cognitive impairment to very severe dementia. Patients in their final stages of disease suffer from complete personality deterioration, incontinence and are dependent on others for basic activities of everyday life. Further criteria for AD are summarized in the Diagnostic and Statistical Manual of Mental Disorders[4], criteria of the National Institute of Neurological and Communicative Disorders and Stroke and the Alzheimer's Disease and Related Disorders Association (now known as the Alzheimer's Association)[5,6]. In spite of that clinical diagnosis is still limited to "probable" or "possible" AD. Unequivocal diagnosis of definite AD continues to require post-mortem histological analysis of the brain.

Histopathological hallmarks of AD are neurofibrillary tangles (NFT) and senile plaques. The plaques are mainly composed of ~4 kDa peptides, the amyloid β peptides that are derived from proteolytic processing of a larger amyloid precursor protein molecule (APP).

Alzheimer's disease is categorised according to its age of onset and/or mode of inheritance. About 1–6% of all AD cases are early onset and are defined as having an age of onset before 65. About 60% of early-onset AD is familial, with 13% inherited in an autosomal- dominant fashion[7,8]. This type of Alzheimer's disease is known as familial Alzheimer's disease (FAD). Patients with FAD may develop symptoms as early as in their

30's or 40's. Most cases of Alzheimer's disease are part of the late onset type, occurring in individuals over 65 years of age, and are sporadic, without a family history.
The overall lifetime risk of any individual to develop dementia is approximately 10–12% but the highest risk factor for AD is advancing age[9,10]. First-degree relatives of a person with AD have a cumulative lifetime risk of developing AD of about 15–30%[11,12]. Disagreement exists as to whether the age of onset of the affected person changes the risk of first-degree relatives[12,13]. The number of additional affected family members probably increases the risk for close relatives, but the magnitude of that increase is unclear.
Sporadic AD is the most common cause of dementia among elderly people and the disease is associated with a significantly higher risk of death compared to other types of dementia[14]. Longitudinal studies provide rates of 10–15 per thousand persons per year for all dementias and 5–8 for AD. Hence, nearly 50% of new dementia cases each year belong to the Alzheimer's type[15,16]. After the age of 65, the risk of acquiring the disease approximately doubles every five years, rising from 3 to 69 per thousand persons per year[15,17]. In 2000, there will be 4.5 million people in the United States suffering from AD. Only 0.3 million people (7%), were between the ages of 65 and 74 years whereas 2.4 million (53%) were between the ages of 75 and 84 years, and 1.8 million (40%) were 85 years of age and older[18]. The charging on the health care system of the United State is estimated to be greater than $100 billion per year, including direct and indirect medical and social service costs[19]. By 2050, the total number of AD patients will increase by almost 3-fold, to 13.2 million and due to the rapid growth of the oldest age groups of the US population, the number of people aged 85 years and older will more than quadruple to 8.0 million[18]. The charging on the health care system can be imagined.

1.2 Heritable causes of Alzheimer's disease

Autosomal dominant gene mutations are potential triggers for AD. Three genes have been identified in which mutations result in early onset familial Alzheimer's disease APP, presenilin 1 (PS-1), and the PS-2 gene.
The APP gene maps to chromosome 21q21.1 and mutations in this gene lead to early onset disease at an age of between 43 and 62 years[20,21]. Mutations in the APP gene might result in altered metabolism of APP, leading to increased production of the Aβ proteins.
The PS-1 locus was identified on chromosome 14q24.3 and mutations in this gene are thought to cause up to 80% of familial Alzheimer's disease cases, with onset between 29 to 62 years of age[22]. PS-1 acts in the γ-secretase complex. However, the exact function of the PS-1 protein is unknown, but it is known to be a transmembrane protein and it is

homologous to SEL-12 in *Caenorhabditis elegans*[23,24]. SEL-12 is known to be involved in cell signaling during development and the PS-1 gene knockout mice reveal skeletal deformations, impaired neurogenesis, and neuronal cell death, leading to death shortly after birth[25,26]. Most PS-1 mutations are gain of function.

The PS-2 Gene locus was identified on chromosome 1q31–42,34[27]. Only two mutations have been identified in the PS-2 gene leading to Alzheimer's disease with an onset between 40 and 88 years of age[21]. The PS-1 and PS-2 protein share 67% homology and are proposed to have a similar function although they are unable to compensate for each other.

Mutations in aforementioned genes lead to an increased production of the 42 amino acid form of Aβ (Aβ$_{1-42}$)[28,29,30,31].

An additional inheritable mutation involved in the pathogenesis of AD occurs in the Apolipoprotein E (ApoE) gene. It exists in 3 allelic forms ε2, ε3 and ε4. The ε2 allele is associated with the lowest late onset Alzheimer's disease (LOAD) risk, whereas ε4 allele increases the risk of developing LOAD 5- to 15-fold[32]. ApoE plays a critical role in regulating brain Aβ peptide levels in the brain. There is evidence that apoE4 enhances Aβ aggregation by increasing the ratio of Aβ$_{1-42}$ to Aβ$_{1-40}$ and reducing Aβ clearance[33,34].

A chromosomal cause for developing AD is found in person with Down Syndrome (DS). Due to the extra copy of the chromosome 21 a lifelong overexpression of the *APP* gene leads to overproduction of Aβ peptides in the brains of DS persons who are trisomic for this chromosome. DS persons develop neuropathologic hallmarks of AD after 40 years[35,36]. Nonetheless, AD remains heterogeneous and complex. The disease does not display a simple mode of inheritance and several genes are known to influence onset and progress of AD.

1.3 Neuritic plaques and Neurofibrillary tangles

The histopathological hallmarks of AD are neuritic plaques and neurofibrillary tangles (NFT), these lesions are not distinctive to AD, and are found in other neurodegenerative disorders as well.

Classic neuritic plaques are spherical structures consisting of a central core of fibrous protein known as amyloid (Aβ) that is surrounded by degenerating or dystrophic nerve ends (neurites). Two types of amyloid-related plaques are recognized in the brains of AD patients:

- o diffuse plaques, which contain poorly defined amyloid but no well-circumscribed amyloid core, and
- o "burnt-out" plaques, which consist of an isolated dense amyloid core.

As mentioned above the amyloid-β contains mainly 40 to 42 amino acid peptides which are derived from proteolytic processing of APP, a type 1 integral membrane protein.

NFT are the other main histopathologic findings in AD. The structure of the NFT was first described by Terry in 1963[37]. 1986 the microtubule associated protein tau (referred to as tau) was determined as the major protein component of NFTs[38]. Tau proteins are expressed predominantly in the axons of neurons in the CNS and peripheral nervous system and it physiological function is to bind and stabilise microtubules[39,40,41]. The activity of Tau as a phosphoprotein is regulated by the balance of phosphorylation and dephosphorylation through different kinases and phosphatases. Key players in this mechanism are GSK3-β as the major tau kinase and PP2A as the major tau phosphatase[42,43,44,45,46,47].

In AD brain there is as much normal tau as in agematched control human brain, but, in addition, the diseased brain contains 4–8-fold of abnormally hyperphosphorylated tau[48,49]. In this state, tau is the major component of the paired helical filaments in NFT[38,50,51]. The intracellular NFTs cause disruption of normal cytoskeletal architecture with subsequent neuronal cell death[52].

Neuritic plaques and neurofibrillary tangles are not distributed evenly across the brain in AD but are concentrated in vulnerable neural systems responsible for learning, memory and survival e.g. the hippocampus.

1.4 Processing of APP

Amyloid-β precursor protein (APP) is a member of a conserved family of type I membrane proteins which in mammals includes also APP like protein 1 (APLP1) and 2 (APLP2). APP and APLP2 are ubiquitous with high expression in neurons, while APLP1 is brain-specific. APP is an important protein that may play a role in recognition of extracellular signals, cell adhesion and apoptosis. In neurons APP is required for synaptogenesis, synapse remodeling and neurite outgrowth[53,54]. There exist three major isoforms of 695, 751, and 770 amino acids all of which are derived from alternative splicing of a single gene product[55] on chromosome 21. In neurons, APP^{695} is the predominantly expressed form and is subject to N- and O-glycosylation within its extracellular/luminal domain. APP^{751} and APP^{770} are expressed mainly in non-neuronal cells of the CNS, especially in glial cells.

During maturation APP gets N-glycosylated in the endoplasmic reticulum and early Golgi. In N-glycosylated state APP is not cleaved by secretases[56]. Further trafficking within the Golgi transforms N-glycosylated APP to O-glycosylated APP and reaches the trans-Golgi network where it enters the secretory pathway[57]. Here two possible APP processing pathways might occur.

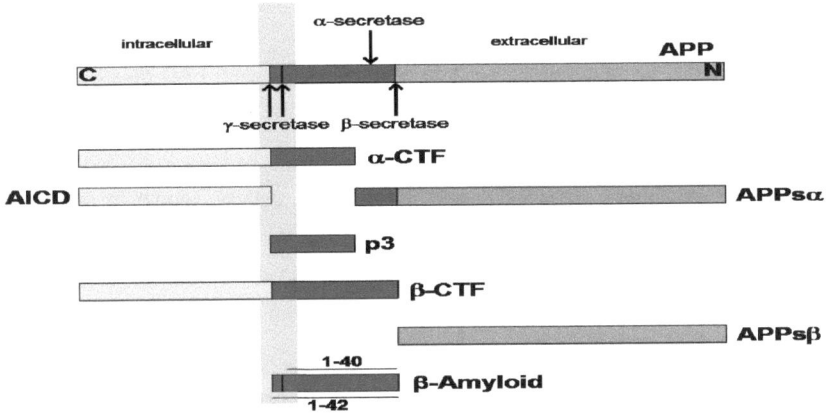

Fig. 1-1 Illustration of APP processing by α-, β- and γ-secretases
Non-amyloidogenic pathway: cleavages of APP by α-secretases produce α-CTFs and APPsα. Further cleavage by γ-secretases produce the p3 stubs and AICD's. Amyloidogenic pathway: processing of APP by β-secretases produce β-CTFs and APPsβ. Cleavage of β-CTFs by γ-secretases produce the Aβ$_{1-40/42}$. Abbreviations: APP, Amyloid precursor protein; CTF, C-terminal fragments; AICD, an intracellular C-terminal domain

a)   Non- amyloidogenic pathway

In the non-amyloidogenic pathway APP is cleaved by the α-secretase, within the Aβ domain between Lys-16 and Leu-17. This is the putative non toxic way of APP processing and generates a 83-amino-acid C-terminal APP fragment (C83) and destroys the Aβ sequence. In addition, a large soluble N-terminal fragment (sAPPα) is released from the cell surface[58]. After subsequently cleavage in the intramembrane region by the γ-secretase a p3 fragment as well as an intracellular C-terminal domain (AICD) is generated[59]. Additional cleavage by caspase-3 between D664 and A665 of AICD produce a 31 aa C-terminal fragment (CTF)[60]. Currently, three members of the ADAM-protein family (a disintegrin and metalloprotease domain) are identified as putative α-secretases. ADAM-9, ADAM-10 and ADAM-17 (also referred as TACE) are suposed to have α-secretase activity allowing to initiate the non-amyloidogenic pathway.

b) amyloidogenic pathway

Amyloidogenic processing of APP requires sequential cleavage by β- and γ-secretase. First APP is cleaved by β-secretase, generating a 99-amino-acid C-terminal APP fragment (C99) and a large soluble N-terminal fragment (sAPPβ). As candidates for β-secretases BACE-1 (β-site APP-cleaving enzyme1) was identified. Cleavage by β-secretase at Asp-1 produces the N-terminus of Aβ peptides. In addition, BACE-1 may also cleave within the Aβ domain at Glu-11, an alternative cleavage site[61]. The Swedish FAD double mutation promotes β-secretase cleavage at Asp-1 and hence increases Aβ production[62]. Further processing of C99 by γ-secretase leads to Aβ-peptide production. γ-secretase is an intramembranous multimeric complex[63] and its cleavage activity seems to be largely nonselective, occurring in at least 3 different sites: Val^{636}, Ala^{638} and Leu^{645} of the APP molecule[64]. The resulting products range in length from 38 to 43 residues but the two major species are $Aβ_{1-40}$ and $Aβ_{1-42}$. The latter is considered to be more amyloidogenic because it was shown to be more prone to aggregate[65].

Notably Aβ peptides are generated only by the cleavage of APP and not by the cleavage of other APP protein family members such as APL-1(amyloid precursor like) in *Caenorhabditis elegans*, Appl (β-amyloid protein precursor like) in *Drosophila,* APP-like protein 1 (APLP1) and APLP2 in mammals, which all lack the Aβ domain[66,67,68,69,70]. The Aβ sequence is the least conserved part between the human and mouse APP sequences. Mouse APP is processed poorly by β-secretase, resulting in approximately threefold lower amounts of Aβ peptide[71]. Therefore in the present study transgenic mice were used carrying the human APP including amino acids exchanges known as the Swedish mutation (APP^{sw}).

1.5 Clearance of Amyloid beta (Aβ)

Clearance of Aβ peptides is achieved by two different pathways: proteolytic degradation, aggregation and receptor-mediated transport from the brain.
A number of different molecules have been implicated in the process of removal of cerebral Aβ by proteolytic degradation. Key players in this mechanism are the enzymes Insulin degrading enzyme (IDE) and Neprilysin (NEP) but additionally, recent data suggest an involvement of endothelin converting enzyme (ECE) in the process of Aβ clearance.

IDE, a 110 kDa zinc metallo-endopeptidase, hydrolyzes several regulatory peptides[72], including insulin, glucagon, atrial natriuretic factor, transforming growth factor α, β-endorphin, amylin, Aβ, and the AICD. IDE is localized in the cytosol, while only a small fraction resides in the plasma membrane. Recent data support a role for IDE in Aβ degradation. Amongst others IDE knockout mice show increased endogenous levels of Aβ and AICD in the brain[73,74]. Chronical overexpression of IDE in APP overexpressing mice diminish the Aβ plaque burden by 50 % and reveals a 50% reduction of soluble and insoluble fraction of $Aβ_{1-40}$ as well as $Aβ_{1-42}$. Furthermore, IDE polymorphismus seems to be associated with late onset AD[75,76]. It is remarkable that IDE is only able to degrade Aβ monomers.

The second mentioned peptidase responsible for the degradation of Aβ is NEP. It is a type II membrane protein and is also referred to as neutral endopeptidase or enkephalinase. NEP, like IDE, hydrolyzes circulating biologically active peptides including enkephalin, neuropeptide Y and others[77]. Intracerebral injections of a lentiviral vector expressing human NEP in a transgenic mouse model of cerebral amyloidosis resulted in a remarkable 50% decrease of cortical amyloid deposits[78]. NEP is localized in the plasmamembrane and owns an extracellularly catalytic site. Therefore NEP is best mounted to be a prime candidate for Aβ degradation on extracellular sites.

Cerebral Aβ is exported across the Blood brain barrier (BBB) via a receptor-mediated transport. The efflux results via low-density lipoprotein receptor-related protein (LRP)[79]. A transport of Aβ via LRP requires initial binding to the LRP ligands apoE and α2 Macroglobulin (α2M). In addition it has been shown that Aβ binds directly to LRP and is transported across the BBB[80,81]. In this case, $Aβ_{1-40}$ is cleared more effectively than the $Aβ_{1-42}$. For that reason $Aβ_{1-42}$ may still require prior binding to the LRP ligand apoE and α2M to be effectively transported out of the CNS.

1.6 Secretases

The processing of APP by the different pathways (amyloidogenic and non-amyloidogenic) requires cleavage by different secretases. In the initial step, α- and β-secretase compete for APP as substrate. These two enzymes cleave at different sites and thereby determine if the amyloidogenic pathway or non-amyloidogenic pathway occurs.

α-secretase

Three members of a disintegrin and metalloprotease domain (ADAM) family have been identified to possess α-secretase activity ADAM-9, ADAM10 and ADAM 17 (also referred to as TACE)[58]. The exact sub-cellular localization of α-secretase remains unclear, however cleavage sites have been proposed to be the trans-Golgi network (TGN) and the cell surface[82,83]. In the CNS ADAM-10 and ADAM-17 are most prominent.

β-secretase

BACE1 (β-site APP-cleaving enzyme1) is essential for initiating Aβ generation and cleaves at the APP Asp-1 residue to form the Aβ N-terminus. BACE1 is an aspartic, type 1 membrane protease with a single transmembrane domain near its C-terminus and a luminal active site that provides an optimal β-secretase site for APP cleavage[61,84,85,86]. Its maximal activity occurs at pH 4.5 and is thus localized within acidic compartments of the secretory pathway[61]. BACE1 is abundant in human cells and its mRNA levels are highest in the brain. Its maximal activity occurs in neurons and to less extend in astrocytes[87]. Like other pepsin family members, BACE1 has two active site motifs and mutation of either causes inactivity[84,88]. Aside from BACE1 there is a homologous molecule BACE2. BACE2 mRNA is expressed at low levels in most human peripheral tissues and at very low or undetectable levels in human brain[61].

γ-secretase

γ-secretase complex is not a single enzyme but requires the interaction of 4 subunits: presenilin (PS), anterior pharynx-defective- 1 (APH-1) , nicastrin, and presenilin enhancer-2 (PEN-2) which are mostly present in a 1:1:1:1 stoichiometry[89].
PS is a polytopic membrane protein consisting of nine trans-membrane-domains (TMD) and pass through an endoproteolytic cleavage that ends in a ~30-kDa N-terminal and ~20-kDa C terminal fragment[90]. This cleavage occurs within the large cytoplasmic loop between TMD6 and TMD7. PS harbors the catalytical active site which is critically required for the aspartyl protease activity of γ-secretase. Apart from the catalytic subunit PS, three other integral membrane proteins, NCT, APH-1, and PEN-2, are essential γ-secretase complex subunits[91,92]. NCT is an ~100-kDa type I membrane glycoprotein with a large ectodomain, a short cytoplasmic domain and recognizes γ-Secretase substrates[93,94]. The other two components, the ~20-kDa seven- TMD protein APH-1 and the smallest subunit, the ~10-kDa hairpin PEN-2 protein, are highly hydrophobic subunits[91,92]. PEN-2 is required for the stabilization of the PS fragments in the complex, whereas the function of APH-1 is currently unclear[95,96].

Presenilin mutations are genetically linked to FAD and increase the production of the aggregation-prone and neurotoxic Aβ_{1-42}.

1.7 IGF-1 signaling cascade

1.7.1 Insulin and Insulin-like-growth factor-1 signaling in Alzheimer's disease

Recent data have implicated insulin and insulin-like growth factor-1 (IGF-1) signaling (IIS) as being involved in the pathogenesis of AD. Current reports suggest that type 2 diabetes mellitus (T2DM) is a risk factor for AD, however, the underlying cellular mechanisms for this association are still unknown[97,98,99,100]. It is conceivable that vascular complications of T2DM result in neurodegeneration[101]. Alternatively, neuronal insulin/IGF-1 resistance might represent the unifying link between T2DM and AD, characterizing AD as a "brain type diabetes"[101,102,103,104]. In agreement with this hypothesis is the observation that insulin receptor (IR) and insulin-like growth factor-1 receptor (IGF-1R) signaling is markedly disturbed in the central nervous system (CNS) of AD patients[105,106,107]. Post mortem investigations of brains from patients with AD revealed a markedly down regulated expression of IR, IGF-1R, and insulin receptor substrate (IRS) proteins[102,108] and these changes progress with severity of neurodegeneration. One common feature in neurons from AD patients is a downregulation of IRS-2 and IGF-1R[102,107]. Other groups reported similar results in AD brains[109]. These findings raise the important question whether changes in IR/IGF-1R signaling (IIS) are cause, consequence, or maybe even compensatory counterregulation of neurodegeneration.

1.7.2 IGF-1 and IGF-1R

IGF-1 is a small molecule of 7500 Da that is found in most tissues. Structurally it is a member of a superfamily of related insulin-like hormones that include IGF-1, IGF-2, insulin and relaxin in vertebrates and bombyxin, locust insulin-related peptide, and molluscan insulin-like peptide in invertebrates[109,110,111,112,113]. A close relative of IGF-1 is insulin with sharing approximately 50% of amino acid homology[114]. IGF-1 is a major growth factor and is involved in proliferation, differentiation, malignant transformation as well as in protection from apotosis. Insulin is predominantly responsible for glucose uptake, food intake and cellular metabolism[115,116,117]. The IGF peptides are single chain polypeptides and derive from a precursor hormone[118]. The final peptide hormone results from processing of the prohormone consisting of A, B, C, and D domains[119]. The gene encoding IGF-1 is highly conserved among mammals, birds and amphibians[118,120,121,122,123,124] and its expression is influenced by hormonal (e.g. growth hormone), nutritional and tissue-specific developmental factors[125,126,127]. The bio-availability of IGFs is regulated by the IGF- binding proteins (IGFBP), a family of six members (IGFBP1 – 6) with a high binding affinity for both IGFs. Thus, they regulate and maintain the biological activity pool of circulating IGF[128]. The IGFBP in turn are regulated by IGFBP proteases which cleave the binding proteins, generating fragments with reduced or no binding affinity for the IGFs[129,130].

The IGF-1 receptor is, like the IR, a member of the ligand-activated receptor tyrosine kinases. Its gene is mapped on chromosome 15 q25-26 consisting of 21 exons spanning over 100kb of genomic DNA. IGF-1 and IR are heterotetrameric trans-membrane glycoproteins consisting of two α- and β-subunits that are covalently linked through disulfide bonds. The α- subunits reside extracellularly with the ligand-binding site and a transmembrane and the cytoplasmatic parts of the receptor are found in the β-subunits[131]. Beside the transmembrane domain is the catalytic subunit with the juxtramembran tyrosine kinase domain located, which link the receptor via the Insulin-receptor-substrates 1-4 (IRS) to the two main downstream signaling cascade, the mitogen activated protein kinase (MAPK) and the phosphatidylinositol 3-kinase (PI3K) cascades[132,133,134]. IGFs and Insulin bind with low affinity to the non-cognate receptor but due to the homology of the IGF-1R and the IR they are able to form hybrid receptors (HR) consisting of IR isoforms and IGF-1R[135]. The HR bind IGF1 with high affinity and insulin with lower affinity, and the relative affinities are dependent on the insulin-receptor isoform that is involved (IRa, IRb)[136]. Their physiological role is unknown but hybrid receptors may be involved in switching signaling from insulin to IGF-1 in certain situation.

1.7.3 IGF-1R/IR signaling

The IGF-1R/IR signaling starts with ligand binding to the corresponding α-subunit of the receptor. This leads to a conformational change and activation of the intrinsic receptor tyrosine kinase followed by intracellular autophosphorylation[137]. For this purpose the ATP-binding site at Lys1003 and the tyrosine kinase domain are required for all functions of the IGF-IR. Trans-phosphorylation between the β-subunits involves $Tyr^{1131, 1135}$ and Tyr^{1136} in the kinase domain and leads to full activation of kinase activity. Phospho-tyrosine residues in specific motifs are docking sites for Src homology 2 (SH2) domain-containing signaling proteins. This phosphotyrosin recognition motif containing proteins,like the IRS-proteins and Shc-proteins, function after being phosphorylated by the receptor tyrosine kinase as adaptor molecules linking the receptor to the PI3K- and MAPK-pathway[138]. IRS, a protein family has at least 4 members (IRS-1 to IRS-4) which are homologous in structure and function but show distinct tissue distribution. IRS-1 and IRS-2 are widely expressed and mediate insulin and IGF-1 action in most tissues including the brain. IRS-3 is largely limited to rodent adipocytes and IRS-4 is primarily and discretely expressed in the brain (hypothalamus) as well as in kidney and thymus[139,140].

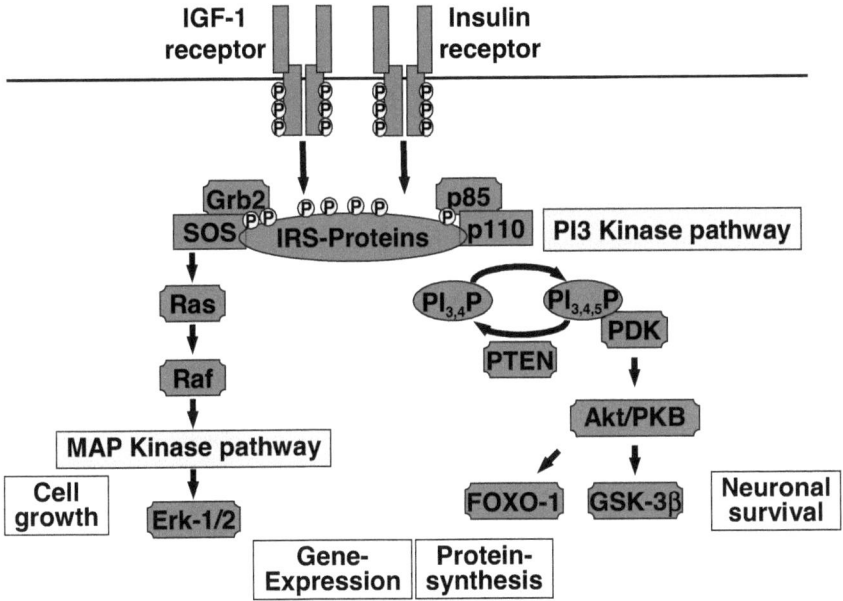

Fig. 1-2 Illustration of IGF-1R/IR signaling cascade
Binding of insulin/IGF-1 ligand to their receptors induce trans-autophosphorylation of the β-subunit and activation of RTK. Activtion leads to recruitment and subsequent phosphorylation of IRS proteins. Phosphorylation of IRS allow for binding of SH-2 domain containing proteins that ends in the activation of MAP- or PI3-kinase signaling pathways. Abbreviations: IGF-1, insulin-like growth factor 1; IRS, insulin receptor substrate; p85/p110, regulatory/catalytic subunit of PI3-kinase; $PI_{3,4}P/PI_{3,4,5}P$, phosphatidylinositol-bi/tri-phosphate; PDK, phosphoinositide-dependent protein kinase; PKB, protein kinase b; GSK3β ,glycogen synthase kinase3β; Grb2, growth factor receptor binding protein2; SOS, son-of-sevenless; Ras, G-protein; c-Raf, proto-oncogene; ERK, extracellular signal-regulated kinase.

The PI3-kinase-pathway: Subsequently phosphorylation of tyrosine residues of IRSs evoke binding of p85 the regulatory subunit of PI3K resulting in the activation of the catalytically active PI3-kinase subunit (p110). This leads to the production of phosphatidylinositol-3,4,5-triphosphate (PIP3) from PIP2 at the plasma membrane, the initial step for activation of several downstream targets, such as the phosphoinositide-dependent protein kinase(PDK)1, AKT (also known as protein kinase B; PKB). AKT phosphorylates glycogen synthase kinase (GSK)3α/β (at $Ser^{21/9}$), which is thereby inhibited, and the forkhead transcription factor FoxO1[141,142]. This step triggers nuclear exclusion of FoxO1 and reduces the expression of its target genes which are involved in oxidative stress protection, apoptosis, metabolism and longevity[143,144]. Notably is that AKT is substrate of mTOR (mammalian target of rapamycin) and therefore mediate signals without activation by IIS. TOR is a Ser/Thr kinase highly conserved from yeast to mammals existing intracellularly in two distinct complexes[145]. When bound to raptor (regulatory-associated protein of mTOR) and GbL (G protein b subunit-like) mTOR regulates protein synthesis, cell growth, proliferation and autophagy in a nutrient- and energy responsive manner. As part of a complex with rictor (rapamycin insensitive companion of mTOR) and GbL, mTOR phosphorylates AKT/PKB and regulates different proteins downstream of AKT. Studies in yeast, *C. elegans*, and *Drosophila* revealed the involvement of TOR in the regulation of life span. In a large-scale screen of single-gene-deletion strains of yeast, mutations in the TOR pathway were associated with an increased life span[146,147]. Thus not only IIS signaling might be responsible for activation of the downstream target of AKT and whose above mentioned impact.

MAP-kinase pathway: The second main pathway that is activated after phosphorylation of the receptors is the MAP-kinase pathway. Phosphorylation of Shc at Tyr950 on IRS proteins leads to the recruitment of Grb2 that binds son-of-sevenless (SOS) which in turn performs nucleotide exchange on Ras[148]. By inducing the exchange of guanosine-diphosphate (GDP) with guanosine triphosphate (GTP), Ras is converted into its active conformation and subsequently recruits c-Raf to the membrane. The increased c-Raf activity is transduced through mitogen-activated protein kinase kinase (MEK) in order to activate extracellular signal-regulated kinase (ERK). ERK regulates transcription factors and thereby influences cell metabolism and proliferation[149,150].

1.8 Mouse Models

In the presented thesis the influence of IGF-1R signaling on the pathogenesis of AD based on a neuronal specific knockout of the IGF-1R (nIGF-1$^{-/-}$) in a transgenic model of AD (Tg2576) has been analyzed. Therefore, the cre/loxP system under the control of the neuron-specific synapsin I promoter has been used.

1.8.1 Conditional IGF-1R knock out (The cre/loxP system)

Originally, gene targeting implicated insertion of an exogenous DNA fragment into an exon critical for target gene function in mouse embryonic stem (ES) cells. The resulting knockout in genes occurs on the basis of homologous recombination[151]. However, mutant gene dysfunction is affected throughout the whole body, often yielding in undesired effects. In contrast to this technique, the Cre/loxP system is able to mediate sitespecific DNA recombination. Originally described in bacteriophage P1 two components are involved, first a 34-bp DNA sequence containing two 13-bp inverted repeats and an asymmetric 8-bp spacer region referred as loxP ("locus of X-over in P1") that targets recombination and second a 343 amino acid monomeric protein called Cre recombinase that mediates the recombination event[152]. Any DNA sequence flanked by two loxP sites is either be excised (loxP sites in same orientation) or inverted (loxP sites in opposite orientation) in the presence of Cre recombinase[153]. The advantages of this system are: i) loxP target sites are small and easily synthesized, ii) no apparent external energy is required[154,155], iii) Cre is a very stable protein and any promoter can drive Cre recombinase expression in the tissue or even celltype of interest. Initiation of gene targeting *in vivo*, using the Cre/loxP system, requires two lines of mice. One mouse line carries the protein of interest flanked by loxP sites ("floxed" gene). These mice should be phenotypically normal because the loxP sites are inserted into introns where they theoretically do not affect gene function. The other mouse line expresses Cre recombinase under the control of a tissue on cell-specific promoters. Cross-breeding of the two mouse lines should result in Cre-mediated gene disruption only in those cells in which the promoter is active.

1.8.2 Cre recombinase expression under the control of the synapsin-1 promotor

Synapsin I (also known as brain protein 4.1), is a neuronal phosphoprotein associated with the membranes of small synaptic vesicles. The synapsin family is composed of synapsin I and synapsin II, which are products of alternative splicing of transcripts from two distinct genes[156]. Two main characteristics distinguish synapsins from most other synaptic vesicle-associated proteins. Firstly they are rather peripheral than integral membrane proteins and secondly they are specific for the nervous system, as there are apparently no homologous proteins in non-neuronal tissues. Injection of synapsin I into Xenopus blastomeres accelerates the structural and functional development of neuromuscular synapses[157,158]. In embryonic hippocampal neurons of synapsin I-deficient mice outgrowth of predendritic neurites and severely retarded axons are observed. Furthermore, synapse formation was significantly delayed indicating that synapsin I plays a role in regulation of axono- and synaptogenesis[159]. Previous use of synapsin-1promoter for Cre recombinase expression determines its activity in cortical and spinal cord neurons but predominantly in the hippocampus[160]. For that reason synapsin I promoter is a good candidate for controlling the cre recombinase to get neuronal specific cre recombinase expression.

1.8.3 The Alzheimer's disease model Tg2576

As mentioned above several mutations affecting APP are capable of inducing FAD. APP^{695SW} mice express transgenic human APP with the two-point mutation (Lys^{670}→Asn, Met^{671}→Leu). This mutation was originally described in a Swedish family suffering from FAD and is therefore called "Swedish" mutation. These mice show age-dependent memory impairments, generally starting in age of 40 weeks of age and several histopathological features, including amyloid plaques, neuritic dystrophy, astrogliosis, reactive microgliosis and to lesser extend abnormal tau phosphorylation[161,162,163]. The 695-amino acid isoform the mutant form of human amyloid precursor protein (APP) was inserted into mice using a hamster prion protein cosmid vector, in which APP^{SW} replaced the prion protein open reading frame. Expression of APP is driven by the hamster prion protein gene promoter. Depending on the genetic background, APP-SW (Tg2576) transgenic mice die early. Since nearly all Tg2576 on a pure C57BL/6 background die within the first months of age it is impossible to investigate amyloid accumulation or IGF-1R signaling in this pure background[164]. Therefore, Taconic APP-SW colony is maintained in a B6/SJL hybrid background. Offsprings of these mice were used in this thesis to investigate APP^{SW} induced lethality, amyloid accumulation as well as IGF-1R signaling during aging in different brain regions[165].

1.9 Aims of this thesis

Recent data show a disturbed Insulin/IGF signaling in patients suffering on AD. To directly address the importance of IGF-1R signaling in the pathogenesis of AD, neuron-specific IGF-1R deleted mice (nIGF-1$^{-/-}$) were crossed with mice expressing the Swedish mutation of human APP695 harbouring the double mutation Lys^{670}→ Asn, Met^{671}→ Leu which was found in a Swedish family with early onset AD (APP^{sw}, Tg2576 mice). Survivals as well as metabolic and somatic factors of the offspring were measured during an observation period of 60 weeks. Biochemical and histophathological analysis of these mice were performed at two different time points to investigate the influence of the neuronal IGF-1R signaling in the pathogenesis of AD.

2 Material and methods

2.1 Chemicals

Acetic acid	Merck, Darmstadt, Germany
Acrylamide / Bis-acrylamide 30%	Rotiphorese® Gel 30 (37.5/1) Carl Roth GmbH + Co. KG, Karlsruhe, Germany
Agarose	Invitrogen Corporation, Carlsbad CA, USA
Aprotinin	Sigma-Aldrich Chemie GmbH, Steinheim, Germany
APS	Ammonium-persulfate AppliChem GmbH, Darmstadt, Germany
AraC	Cytosine arabinoside, Sigma-Aldrich Chemie GmbH, Steinheim, Germany
Avertin	Sigma-Aldrich Chemie GmbH, Steinheim, Germany
Benzamidine	Sigma-Aldrich Chemie GmbH, Steinheim, Germany
β-mercaptoethanol	Sigma-Aldrich Chemie GmbH, Steinheim, Germany
Bradford reagent	Bio-Rad Laboratories GmbH; Germany
Bromophenol blue	AppliChem GmbH, Darmstadt, Germany
BSA > 96 %	Bovine serum albumin Sigma-Aldrich Chemie GmbH, Steinheim, Germany
Desoxy-Ribonucleotid-Triphosphate (dNTPs)	Fermentas GmbH, St. Leon-Rot, Germany
DMSO	Dimethyl sulfoxide Sigma-Aldrich Chemie GmbH, Steinheim, Germany
DNase	Roche, Mannheim, Germany
DTT	Dithiothreitol AppliChem GmbH, Darmstadt, Germany
EDTA	Ethylenediaminetetraacetic acid AppliChem GmbH, Darmstadt, Germany
Ethanol	AppliChem GmbH, Darmstadt, Germany
Ethidium bromide	Sigma-Aldrich Chemie GmbH, Steinheim, Germany
Glycerol	Glycerin, AppliChem GmbH, Darmstadt, Germany

Glycine	AppliChem GmbH, Darmstadt, Germany
HEPES	Sigma-Aldrich Chemie GmbH, Steinheim, Germany
IGF	Sigma-Aldrich Chemie GmbH, Steinheim, Germany
Isopropanol	AppliChem GmbH, Darmstadt, Germany
Kaiser's glycerol gelatine	Merck, Darmstadt, Germany
KCl	potassium chloride, Sigma-Aldrich Chemie GmbH, Steinheim, Germany
Methanol 99%	Carl Roth GmbH + Co. KG, Karlsruhe, Germany
Magnesium chloride	Merck, Darmstadt, Germany
NP-40	Polyglycol ether (Nonidet® P40 Substitute) FLUKA Chemika/Biochemika Chemie AG, Buchs, Switzerland
PMSF	Phenylmethylsulphonylfluoride Sigma-Aldrich Chemie GmbH, Steinheim, Germany
Potassium hexacyanoferrat II	Merck, Darmstadt, Germany
Potassium hexacyanoferrat III	Merck, Darmstadt, Germany
Proteinase K	Roche, Mannheim, Germany
SDS	Sodium dodecyl sulfate AppliChem GmbH, Darmstadt, Germany
Sodium bicarbonate	Carl Roth GmbH + Co. KG, Karlsruhe, Germany
Sodium chloride	Carl Roth GmbH + Co. KG, Karlsruhe, Germany
Sodium orthovanadate	Sigma-Aldrich Chemie GmbH, Steinheim, Germany
TEMED	N,N,N',N'-Tetramethylethylenediamine Sigma-Aldrich Chemie GmbH, Steinheim, Germany
Thioflavin S	Sigma-Aldrich Chemie GmbH, Steinheim, Germany
Tris	AppliChem GmbH, Darmstadt, Germany
TritonX-100	AppliChem GmbH, Darmstadt, Germany
Trypsin	Roche, Mannheim, Germany

TWEEN 20®	Polyoxyethylene (20) sorbitan monolaurate, Caesar and Lorentz GmbH, Bonn, Germany
X-gal	PEQLAB Biotechnologie GmbH, Erlangen, Germany
Xylol	AppliChem GmbH, Darmstadt, Germany

2.1.1 Buffer and solution

BME	Basal medium eagle, Invitrogen Corporation, Carlsbad CA, USA
Cell lysis buffer	150 mM NaCl 50 mM Tris-HCl (pH 7.4) 5 mM EDTA 1 % Nonidet® P40 Substitute
Organ lysis buffer	50 mM HEPES (pH 7.4) 50 mM NaCl 1 % Triton X-100 10 mM EDTA 0.1 M NaF 17 µg/ml Aprotinine 2 mM Benzanidine 0.1 % SDS 1 mM Phenylmethylsulfonyl fluoride (PMSF) 10 mM Na_3VO_4
HBSS	Hanks' balanced salt solution, Invitrogen Corporation, Carlsbad CA, USA
SDS-PAGE running buffer	194 mM Glycine 25 mM Tris 0.1 % SDS
4 x SDS sample buffer	250 mM Tris-HCl (pH 6.8) 200 mM DTT 40 % Glycerol 8 % SDS 0.01 % Bromophenol blue
Stripping solution	62.5 mM Tris-HCL pH 6.8 100 mM β-mercaptoethanol 2% SDS

TBS buffer (pH 7.6)	137 mM NaCl 20 mM Tris
TBS-T buffer (pH 7.6)	137 mM NaCl 20 mM Tris 0.1 % Tween 20®
Western Blot antibody solution	137 mM NaCl 20 mM Tris 5 % Western Blocking Reagent (Roche)
Western Blot blocking solution	137 mM NaCl 20 mM Tris 10 % Western Blocking Reagent (Roche)
Western Blot transfer buffer	194 mM Glycin 25 mM Tris 20 % Methanol (99%) 0.05 % SDS

– ECL; Amersham ECLTM Western Blotting Detection Reagents, GE Healthcare UK Ltd; England

– Fetal bovine serum (FBS); Invitrogen GmbH; Germany

– Pen/Strep; Penicillin Streptomycin (P/S); 10,000 Units/ml Penicillin, 10,000 µg/ml Streptomycin; Invitrogen GmbH; Germany

– Phosphate buffered saline 10 fold (pH 7.2); Invitrogen GmbH; Germany

– Protein Standard Ladder; Precision Plus Protein Kaleidoscope Standards; Bio-Rad Laboratories GmbH; Germany

– Trypsin; Roche, Mannheim, Germany

– Western Blocking Reagent; Roche Diagnostics GmbH; Germany

2.1.2 Kits

– α-Secretase Activity Kit	R&D Systems, Inc., USA; Catalog # FP001
– β-Secretase Activity Kit	R&D Systems, Inc., USA; Catalog # FP002
ELISA Aβ$_{1-40}$	Invitrogen Corporation, Carlsbad CA, USA Cat# KHB3481
ELISA Aβ$_{1-42}$	Invitrogen Corporation, Carlsbad CA, USA Cat# KHB3441

2.1.3 Primary Antibodies

– Actin Antibody; Monoclonal mouse antibody raised against an epitope conserved in human actin; MP Biomedicals, USA; Item # 69100; Western Blotting Dilution 1:5000

– ADAM 10 Antibody; Polyclonal rabbit antibody raised against human ADAM10 (H-300); Santa Cruz Biotechnology, Inc., USA; Item # sc-25578; Western Blotting Dilution 1:1000

– ADAM 17/TACE Antibody; Polyclonal rabbit antibody raised against human ADAM17/ TACE; Assay Designs, Inc., USA; Item # 905249; Western Blotting Dilution 1:1000

– AKT Antibody; Polyclonal rabbit antibody raised against endogenous levels of total AKT1, AKT2 and AKT3 proteins; Cell Signaling Technology, Inc., USA; Item # 9272; Western Blotting Dilution 1:1000.

- ApoE Antibody; Polyclonal goat antibody raised against a peptide mapping the C-terminus of apoE of mouse origin; Santa Cruz Biotechnology, Inc., USA; Item # sc-6384; Western Blotting Dilution 1:1000

– APP C-Term (Amyloid Precursor Protein, C-Term) Antibody; Synthetic peptide developed in rabbit raised against the C-terminal of human APP 695 (amino acids 676-695); Sigma-Aldrich, USA; Item # A8717; Western Blotting Dilution 1:1000

- α2M Antibody Polyclonal goat antibody raised against epitope mapping near the N-terminus of α-2M of human origin Santa Cruz Biotechnology, Inc., USA; Item # sc-8513; Western Blotting Dilution 1:1000

– BACE-1 (Beta Site APP Cleaving Enzyme 1) Antibody; Polyclonal rabbit antibody raised against amino acids 458 to 501 of human BACE; Chemicon (Millipore), USA; Item # AB 5832; Western Blotting Dilution 1:1000

– Beta Amyloid Antibody; Polyclonal rabbit antibody raised against several isoforms of β-amyloid peptide (Aβ), such as Aβ1-40, Aβ1-42 etc, regardless of phosphorylation state; Cell Signaling Technology, Inc., USA; Item # 2454; Western Blotting Dilution 1:1000

– Erk Antibody; Polyclonal rabbit antibody raised against endogenous levels of total p44/42 MAP kinase (Erk1/Erk2) protein; Cell Signaling Technology, Inc., USA; Item # 9102; Western Blotting Dilution 1:1000

- Foxo1 Antibody; Polyclonal rabbit antibody raised against epitope corresponding to amino acids 471-598 of FKHR of human origin Santa Cruz Biotechnology, Inc., USA; Item # sc-11350; Western Blotting Dilution 1:1000

– GSK-3-β Antibody; Monoclonal rabbit antibody raised against endogenous levels of total GSK-3β protein; Cell Signaling Technology, Inc., USA; Item # 9315; Western Blotting Dilution 1:1000

– Holo APP Antibody; Polyclonal rabbit antibody raised against endogenous levels of several isoforms of both mature and immature amyloid β (A4) precursor protein, including APP695, APP770 and APP751; Cell Signaling Technology, Inc., USA; Item # 2452; Western Blotting Dilution 1:1000

- IDE Antibody; Polyclonal rabbit; Millipore Corporation 290 Concord Road, Billerica, MA 01821, USA; Item # AB9210; Western Blotting Dilution 1:1000

– IGF-1 Receptor β Antibody; Polyclonal rabbit antibody raised against endogenous levels of IGF-IR β. Does not cross-react with insulin receptor; Cell Signaling Technology, Inc., USA; Item # 3027; Western Blotting Dilution 1:1000

– IR-β Antibody; Polyclonal rabbit antibody raised against a peptide mapping at the Cterminus of insulin Rβ (C19) of human origin; Santa Cruz Biotechnology, Inc., USA; Item # sc-711; Western Blotting Dilution 1:1000

– IRS-1 Antibody; Monoclonal rabbit antibody raised against C-terminal 14 amino acid peptide ([C]YASINFQKQPEDRQ) of rat liver IRS-1. Rat, mouse and human crossreactivity; Upstate Cell Signaling Solutions, USA; Catalog # 06-248; Western Blotting Dilution 1:1000

– IRS-2 Antibody; Polyclonal rabbit antibody raised against endogenous levels of total IRS-2 protein; Cell Signaling Technology, Inc., USA; Item # 4502; Western Blotting Dilution 1:1000

- Nepriliysin Antibody; Polyclonal rabbit; Millipore Corporation 290 Concord Road, Billerica, MA 01821, USA; Item # AB5458; Western Blotting Dilution 1:1000

– Phospho-AKT Antibody; Polyclonal rabbit antibody raised against endogenous levels of AKT1 only when phosphorylated at Ser473. Also recognizes AKT2 and AKT3 when phosphorylated at the corresponding residues; Cell Signaling Technology, Inc., USA; Item # 9271; Western Blotting Dilution 1:1000

– Phospho-p44/42 MAP Kinase (Thr202/Tyr204) Antibody; Polyclonal rabbit antibody raised against endogenous levels of p44 and p42 MAP Kinase (Erk1 and Erk2) when phosphorylated either individually or dually at Thr202 and Tyr204 of Erk1 (Thr185 and Tyr187 of Erk2); Cell Signaling Technology, Inc., USA; Item # 9101; Western Blotting Dilution 1:1000

– Phospho-GSK-3β (Ser9) Antibody; Polyclonal rabbit antibody raised against endogenous levels of GSK-3β only when phosphorylated at serine 9; Cell Signaling Technology, Inc., USA; Item # 9336; Western Blotting Dilution 1:1000

– Phospho-GSK-3α/β (Ser21)/(Ser9) Antibody; Polyclonal rabbit antibody raised against endogenous levels of GSK-3α/β only when phosphorylated at serine 21 or 9; Cell Signaling Technology, Inc., USA; Item # 9327; Western Blotting Dilution 1:1000

- Phospho-Foxo1 (Ser 256) Antibody Polyclonal rabbit Antibody detects endogenous levels of Fox01 only when phosphorylated at serine 256 Cell Signaling Technology, Inc., USA; Item # 9336; Western Blotting Dilution 1:1000

− Presenelin 1 (C20) Antibody; Polyclonal goat antibody raised against a peptide mapping at the C-terminus of Presenilin 1 of human origin; Santa Cruz Biotechnology, Inc., USA; Item # sc-1244; Western Blotting Dilution 1:1000

- PTEN Polyclonal Rabbit mAb detects endogenous levels of total PTEN protein; Cell Signaling Technology, Inc., USA; Item # 138G6; Western Blotting Dilution 1:1000

2.1.4 Secondary Antibodies

− Anti Goat IgG (whole molecule), peroxidase conjugated; Affinity isolated antigen specific antibody obtained from rabbit anti-goat antiserum by immunospecific purification; Sigma-Aldrich, USA; Item # A5420; Western Blotting Dilution 1:1000

− Anti Mouse IgG (Fab specific), peroxidase conjugated; Developed in goat using purified mouse IgG Fab fragment as immunogen, the antibody is isolated from goat anti-mouse IgG antiserum by immunospecific purification; Sigma-Aldrich, USA; Item # A9917; Western Blotting Dilution 1:15000

− Anti Rabbit IgG, peroxidase conjugated; Developed in goat using purified rabbit IgG as immunogen, the antibody is isolated from goat anti-rabbit IgG antiserum by immunospecific purification; Sigma-Aldrich, USA; Item # A6154; Western Blotting Dilution 1:1000

2.2 Materials

Blotting chamber Trans-Blot® Semi-Dry Transfer Cell
Bio-Rad Laboratories, USA

Blotting membrane Immun-BlotTM PVDF Membrane for Protein Blotting
Bio-Rad Laboratories, USA

Blotting paper Whatman® Gel Blotting Paper
Schleicher & Schuell, Germany

Cover-slips Cover glasses 24 x 50 mm
VWR International GmbH, Germany

6-well culture dishes poly-L-lysine coated
Nunc – Thermo Fisher Scientific, Denmark

iCycler Thermocycler
Bio-Rad Laboratories, USA

Gewebe-Homogenisator
VWR International GmbH, Germany

Microplate reader Mithras LB 940 multimode microplate reader
Berthold Technologies GmbH & Co. KG, Germany

Microscope Fluorescence Microscope Eclipse E800
Nikon Instech Co., Ltd. Kanagawa, Japan

Microscope slides Microscope slides 76x26 mm
Menzel GmbH &Co KG, Braunschweig, Germany

Minigel-Twin Gel Electrophoresis Apparatus, Minigel-Twin
Biometra GmbH, Germany

Nanodrop NanodropTM Spectrophotometer ND 1000
ThermoFisher Scientific, USA

NMR Analyzer minispec mq7.5
Burker Optik, Ettlingen, Germany

Photo-paper Amersham HyperfilmTM ECL
GE Healthcare UK Ltd, England;

Powerpac Biometra Standard Power Pack P25
Biometra GmbH, Germany

Thermomixer
Eppendorf, Hamburg, Germany

2.3 Methods

2.3.1 Isolation of genomic DNA

Mouse tail biopsies were incubated o/n in lysis buffer (100 mM Tris HCl (pH 8.5), 5 mM EDTA, 0.2% (w/v) SDS, 0.2M NaCl, 500 mg/ml proteinase K) in a thermomixer at 55°C. DNA was then precipitated from solution by adding an equivalent of isopropanol. After centrifugation (13.000 rpm, 15 minutes, RT) supernatant were removed. Subsequent to adding of 150 µl of 70% ethanol samples were centrifuge a second time (13.000 rpm, 15 minutes, RT). Afterwards DNA pellet was dried and resuspended in 50 µl double distilled water (ddH$_2$O).

2.3.2 Quantification of Nucleic acid

DNA concentration was measured at 260nm with Nanodrop® ND-100 UV Spectrophotometer.

2.3.3 Polymerase Chain Reaction (PCR)

The PCR method was used to genotype mice for the presence of floxed alleles or transgenic expression of APP or synCre with primers listed in Table 2-1 reaction was performed in a Thermocycler PCR machine. All amplifications were performed in a total reaction volume of 50 µl, containing a minimum of 100 ng template DNA, 25 pmol of each primer, 25 µM dNTP Mix, 1 x goTaq reaction buffer and 1unit of goTaq DNA polymerase. Standart PCR programs started with 4 minutes denaturation at 95°C, followed by 30 – 45 cycles consisting of denaturation at 95°C for 45 seconds, annealing at oligonucleotide-specific temperatures for 30 seconds and elongation at 72°C for 30 seconds and final elongation step at 72 °C for 7 minutes

Primer	Sequences 5'-3'	T$_{Annealing}$ (°C)	Oriantation
SynCre 5'	ACCTGAAGATGTTCGCGATTATCT	57	sense
SynCre 3'	ACCGTCAGTACGTGAGATATCTT	57	antisense
Tg2576 5'	CTGACCACTCGACCAGGTTCTGGG	66	sense
Tg2576 3'	GTGGATAACCCCTCCCCCAGCCTAGACCA	66	antisense
IGF-1R 5'	TCCCTCAGGCTTCATCCGCAA	59	sense
IGF-1R 3'	CTTCAGCTTTGCAGGTGCACG	59	antisense

Table 2-1 Oligonucleotides used for genotyping
PCR-amplified DNA fragments were applied to 2% (w/v) agarose gels (1 x TAE, 0.5 µg/ml ethidium bromide) and electrophoresed at 150 V.

2.3.4 Animals, breeding and genotyping

IGF1Rlox/lox mice were generated as described above and crossed with Synapsin-Cre (synCre) mice to achieve neuron-specific deletion. Mice which did not express APPsw or synCre served as controls. Animals were housed in a 12-h light/dark cycle (07:00 on/19:00 off) and were fed a standard rodent diet. Tg2576 mice with transgenic expression of the Swedish mutation of APP695 (APP^{SW} were purchased from Taconic Corporate, Hudson, NY, USA) in a B6/SJL background. Since the genetic background of Tg2576 mice might influence mortality[164,166], we used the APP^{SW} model from taconic in a B6/SJL background and crossed these mice back for 3 generations in a C57BL/6 background. Due to this approach similar mortality rates of Tg2576 mice were obtained as described in the literature[167,168]. All animal procedures were performed in accordance with the German Laws for Animal Protection and were approved by the local animal care committee and the Bezirksregierung Köln.

2.3.5 Histology and immunostaining

X-gal staining

SynCre mice were crossed with RosaArte1 reporter mice (28). SynCre-LacZ mice were anesthetized and transcardially perfused with physiologic saline solution followed by 4% paraformaldehyde (PFA) in 0.1 M phosphate-buffered saline (PBS; pH 7.4). Brains were then frozen in tissue-freezing medium (Jung Tissue Freezing Medium; Leica Microsystems, Wetzlar, Germany) and sectioned on a cryostat. Slides containing sagitally dissected brains were fixed 15 minutes with ice cold methanol at -20°C following by 2 x washing in PBS. Then the slides were incubated in X-gal staining solution (5 mM potassium hexacyanoferrat II, 5mM Potassium hexacyanoferrat III, 2 mM $MgCl_2$ and 1 mg/ml X-gal dissolved in DMSO) over night at 37°C light protected. Next the slide were washed 3 x with PBS and 1 x in distilled H_2O. Afterwards the slides were mounted in Kaiser's glycerol gelatine and stored light protected at 4°C.

Thioflavin-S staining

Tg2576 and nIGF-1R$^{-/-}$ mice were anesthetized and transcardially perfused with physiologic saline solution followed by 4% paraformaldehyde (PFA) in 0.1 M phosphate-buffered saline (PBS; pH 7.4). Brains were then fixed for 48h in 4% PFA solution at 4°C. Afterwards brains were paraffin embedded and sectioned. Paraffin slides were sequently incubated in 2 x Xylol for 15 minutes and in a decent order 2 x in 100%, 1 x 96%, 1 x 70% ethanol for 1 minute. Then the slides were washed for 10 minutes in distilled water. All followed steps were performed light protected. Slides were incubated in 0,1% Thioflavin-S staining solution for 3 minutes. Subsequent they were washed 3 x in distilled water followed by a differentiation step with 1% acetic acid for 20 minutes. Afterwards slides were washed with normal H_2O, mounted in Kaiser's glycerol gelatine and stored light protected at room temperature.

2.3.6 Metabolic characterization, glucose, and insulin tolerance tests

Mice were weighed weekly beginning at weaning in week 4 until performance of glucose and insulin tolerance tests in weeks 10 and 11. From week 12 blood glucose and weight was measured every 4 weeks.

For insulin tolerance tests animals were starved overnight (16 h) and injected with 0.75 U/kg body weight of human insulin (Novo Nordisk, Copenhagen, Denmark) into the peritoneal cavity. Blood glucose values were measured in blood collected from the tail tip immediately before and 15, 30, and 60 min after the injection. Blood glucose measurements were performed using a blood glucose meter (GlucoMen, A. Menarini diagnostics, Berlin-Chemie, Neuss, Germany). Results were expressed as percentage of initial blood glucose concentration.

For glucose tolerance tests mice were starved overnight (16 h). Animals were injected with 2 g/kg body weight of glucose into the peritoneal cavity. Glucose levels were determined in blood collected from the tail tip immediately before and 15, 30, 60, and 120 minutes after the injection using a glucose meter.

2.3.7 Analysis of Body composition

Nuclear magnetic resonance (NMR) was employed to determine whole body composition of live animals using the NMR Analyzer minispec mq7.5. Radiofrequency (RF) pulse sequences are transmitted into the tissue. In response, RF signals are generated by the hydrogen in the tissue, which are detected by the minispec. The amplitude and duration of these signals are related to properties of the material.

2.3.8 Isolation of cerebellar granule cells

Cerebellar granule neurons were isolated from 5 days old mouse litters. All manipulations were performed at 4°C unless indicate d otherwise. Individual cerebella were isolated, the meninges were removed using a dissecting microscope and the cerebella were washed three times in HHGN (1x HBSS, 2.5 mM Hepes, pH 7.4, 35 mM glucose, 4 mM sodium bicarbonate). Cerebella were then incubated in trypsin solution (10 mg/ml of trypsin, 100 μg/ml DNase, in HHGN, pH 7.0 with 0.1 N NaOH) for 15 min. at room temperature. Cerebella were placed on ice, washed three times in HHGN and then triturated ~25 times with 1ml of DNAse solution (10 μg/ml of DNase in basal medium eagle (BME)). The cells were allowed to settle for 5 min at room temperature, the supernatant was transferred into fresh tubes and the remaining pellet was triturated with an additional 1 ml of DNAse solution for another 25 times. After settling, the supernatants were combined and the cells were centrifuged for 5 minutes at 1000 x g. Cell pellets were suspended in BME containing

10% fetal calf bovine, 100 U of penicillin-streptomycin, 2 mM gluthamine and 25 mM KCl (culture medium), counted and plated on poly-D-lysine coated 6-well culture dishes. After 24h 10 µM cytosine arabinoside (araC) was added to cultures to inhibit proliferation of non-neuronal cells. Cells were cultivated for 10 days.

2.3.9 Immunoblotting

Brain regions were lysed in buffer (50 mM HEPES (pH 7.4), 50 mM NaCl, 1 % Triton X-100, 10 mM EDTA, 0.1 M NaF, 17 µg/ml Aprotinine, 2 mM Benzanidine, 0.1 % SDS, 1 mM Phenylmethylsulphonyl fluoride (PMSF) and 10 mM Na_3VO_4) using a dounce hand homogenizer. Protein expression was determined from brain region lysates (50-100 µg) dissolved in Laemmli buffer and resolved on 10 % or 15 % SDS-PAGE.

2.3.10 Gel Electrophoresis

Sodium dodecyl sulfate (SDS) polyacrylamide gel electrophoresis (PAGE), a technique referred to as SDS-PAGE, is used to separate proteins based on their molecular size. The negatively charged, anionic detergent SDS binds to heat denatured, linearized proteins within a given sample and applies a negative charge to each protein in proportion to its mass. The samples are subsequently transferred to one end of a layered polyacrylamide gel. The gel is located between glass plates and mounted into a gel apparatus (Minigel-Twin). By applying an electric current to the gel-matrix that is submerged in a buffer solution, the negatively-charged proteins migrate throughout the gel in a size depending manner: Short proteins will fit more easily through the gel matrix and hence travel longer distances, whereas larger ones are hampered in locomotion and will cover shorter distances respectively. The SDS-PAGE gels used were heterogeneous, consisting of a large pore stacking gel and a small pore resolving gel. The stacking gel served to gather SDS-coated proteins. These were concentrated to several folds in a thin starting zone, before entering the resolving gel where proteins were ultimately separated. Resolving gels were used in concentrations of either 10 or 15 % acrylamide contingent depending on sizes of the sought after proteins. Gels were poured between glass plates which were held apart by spacers. A comb was applied into the stacking gel to create a number of gel pockets in which protein samples were pipetted.

Reagents	Stacking gel (5%)	Resolving gel (10%)	Resolving gel (15%)
dd H_2O	2.74 ml	6.34 ml	3.5ml
Acrylamid (30%)	680 µl	4 ml	5.25 ml
Tris 1,5M pH 8,8		1,5 ml	1.3 ml
Tris 1 M pH 6,8	500 µl		
SDS (10%)	40 µl	120 µl	105 µl
APS (10%)	80 µl	160 µl	140 µl
TEMED	4 µl	12 µl	10.5 µl

Table 2-2 SDS-PAGE mini gels (2 x)

Polyacrylamide gels must be carefully polymerized by the mixing of the appropriate salts and buffers, monomeric units of acrylamide, an initiator of polymerization and a catalyst. Ammonium-persulfate (APS) initiates gel polymerization. N,N,N',N'-tetramethylethyl-enediamine (TEMED) is the catalyst and must be added last just before the gel is poured. *Sample preparation:* After determining the protein concentration of the cell lysates, SDS-PAGE samples were prepared. These consisted of 50 µg of protein and the corresponding amount of 4 x SDS sample buffer. Just before being loaded into the gel, samples were boiled at 95°C for 5 min. and centrifuged at 13.000 rpm for 2 min. In order to determine relative molecular weight of the proteins of 10 µl of molecular weight marker was loaded into the first and the last slot of every gel. Electrophoresis of the stacking gel was performed at 100 Volt, resolving gel at 150 Volt.

2.3.11 Western Blot

Western blot is defined as transferring electrophoretically separated proteins from a gel matrix onto a polyvinylidene difluoride (PDVF) or nitrocellulose membrane. In a second step the membrane is incubated in a solution containing antibodies to detect a protein of interest. In our lab the transfer process was performed by a method referred to as semidry-blotting. It relies upon an electric current to drive SDS-coated proteins from within the gel onto a PVDF membrane while maintaining inter-spatial protein organization. The use of this technique requires a semi-dry blotting chamber in which the resolving gel and the PVDF membrane are sandwiched between sheets of buffersoaked filter paper. This stack is placed between two plate electrodes in a horizontal configuration, resembling anode and cathode from top to bottom. The plate electrodes are separated only by the stack of filter paper, providing high field strength (V/cm) across the gel.

In detail the procedure was performed as described below: Following gel electrophoresis, the gel was removed from in between the glass plates. Using a gel knife, the top stacking parts as well as the bottom margin of the gel were trimmed away. In a clean container 7 sheets of Whatman's filter paper (7 cm x 9 cm in size) were soaked in transfer buffer. 4 of these pre-soaked pieces were placed on the cathode plate of the blotting chamber. A piece of PVDF membrane (6.5 cm x 8.5 cm in size) was prewetted using 99% methanol for 10 seconds and placed on top of the four layers of filter paper. Subsequently the trimmed resolving gel was placed upon the PVDF membrane and covered by the 3 remaining pieces of filter paper. Air bubbles were removed by gently rolling a glass pipette over the stack. Finally the blotting chamber was closed by placing the anode plate on the stack. This was done carefully without disturbing the stack structure. The transfer was performed using an electric current of 200 milli-amperes (mA), transfer time was set to 60 minutes. For target proteins of 100 kDa and more in size, transfer time was adjusted to 90 minutes. Prior to incubating the PVDF membrane with antibodies for detection of the target proteins, the membrane was immersed in a "blocking" solution at room temperature for 60 minutes. The solution was composed of tris buffered saline (TBS) with 10 % Western Blocking Reagent. The procedure was performed to assure saturation of vacant membrane protein binding sites. By preventing non-specific antibody binding to the membrane, background staining was reduced. In the subsequent detection process, the membrane was incubated with antibodies using a two step procedure. The primary antibody, raised against a protein of interest, was applied over night (12-16h) at 4° Celsius on a rocker. The antibody solution consisted of TBS, 5 % Western Blocking Reagent and the diluted antibody. Following the next morning the membrane was thoroughly washed 5 times for 15 minutes to remove unbound antibodies. The washing process was performed at room temperature on a rocker using a solution of TBS containing 0.1 % TWEEN 20® (TBS-T). The secondary antibody was subsequently applied for 60 minutes. It was directed at a constant portion of the primary antibody. The secondary antibody was conjugated to horseradish peroxidase (HRP) by protein cross-linking. After the 60 minute incubation time, the membrane was washed 5 times for 5 minutes using TBS-T. Using an enhanced chemiluminescence assay target proteins were detected by photographic film.

For final analysis concentrations of each target protein were determined twice utilizing discrete PVDF membranes that derived from independent cell-lysate samples. To assess potential error arising from deviations in the employed amount of gross protein we applied β-actin-specific antibodies to every membrane in order recognize differences in protein loading.

Enhanced Chemiluminescence Assay

The enhanced chemiluminescence (ECL) assay is a light-emitting system designed to detect membrane bound proteins. It is based upon horseradish peroxidase (HRP) that is conjugated to a secondary antibody, and on the ECL substrate luminol. HRP catalyzes the oxidation of luminol which then emits light. The light is chemically enhanced and recorded on film for further analysis by densitometry. The membrane was soaked in the detection reagent for two minutes (Amersham ECL™ Western Blotting Detection Reagent). It was subsequently covered by transparent plastic foil and placed in a metal cassette. In the darkroom the membrane was exposed to photosensitive film (AmershamTM Hyperfilm ECL) Depending on the intensity of the membrane emitted light, film-exposure times varied between 10 seconds and 30 minutes. Following this, the film was developed.

Membrane Stripping

In cases where the PVDF membrane was reprobed to detect a different protein of interest, the membrane was "stripped" to clear all previously bound antibodies prior to being incubated with antibodies for a second time. In order to do so, the membrane was incubated in stripping solution (62.5 mM Tris-HCl pH 6.8, 2% SDS and 100 mM β- mercaptoethanol) for 20 minutes at 55°Celsius in a water-bath. Following this, the membrane was thoroughly washed 5 times for 10 minutes using TBS-T and blocked for 60 minutes using standard procedure. Thereafter the membrane was ready to be reprobed with primary antibody.

2.3.12 Secretase Actvity Assays

In order to detect the enzymatic activity of α- and β-secretase, classes of proteases that are associated with the cleavage of APP, we performed a number of secretase activity assays. By adding secretase-specific, reporter conjugated peptides to cell lysates, secretase activity is determined using a fluorometric reaction. This reaction is based upon cleavage of the reporter conjugated peptides by secretase, resulting in the release of a fluorescence signal. The fluorometric reaction is proportional to the level of enzymatic activity in the cell lysate. Fresh dissected brain regions were homogenized in cold extraction buffer provided by the kit, followed by 45 min incubation on rotator wheel at 4°C. Protein concent rations were determined by bradford protein assay. Samples were prepared in triplicates. Each sample contained 100 µg of brain lysate protein that was diluted in 50 µl of cold cell extraction buffer (provided by the kit), 50 µl of cold reaction buffer (provided by the kit) and 5 µl of prewarmed (20°C) substrate. The constituents were s equentially pipetted into wells of 96 a well micro-plate. The microplate was covered with foil, tapped gently to mix and incubated in the dark at 37°C for 60 min. Sampl es were analyzed in a fluorescence micro-plate reader using a light filter that allowed excitation between 335 and 355 nanometer (nm) wavelength. Collection of emitted light was accomplished at 580 nm wavelength (Emission-Filter f 535; Lamp Energy 3500). For data analysis two negative control samples were included; one without brain lysate, and one neither containing brain lysate nor substrate.

2.3.13 ELISA β-Amyloid$_{1-40/42}$

Amyloid was extracted using 5 M guanidine HCl in 50 mM Tris HCl, pH 8.0. Then ELISAs of βA1-40/1-42 were performed following the manufacturers protocol (Cat# KHB3481/ 3441, Invitrogen Corporation, Carlsbad, CA, USA)

2.3.14 Statistical analysis

To quantify the changes in optical density we used the software AIDA (Version 4.00.027, Raytest, Straubenhardt, Germany). For statistical analysis of the different study groups unpaired Student's t-test was performed. Statistical significance was defined as *p<0.05. For Kaplan Meier analysis the XLSTAT-Life software, a Microsoft Excel add-in (www.xlstat.com) was used. For comparison of the different study groups Wilcoxon rank tests were performed. Statistical significance was defined as minimum *p<0.05.

3 Results

Growing evidence indicates insulin and insulin-like growth factor-1 (IGF-1) signaling (IIS) as being involved in the pathogenesis of AD. Insulin receptor (IR) and insulin-like growth factor-1 receptor (IGF-1R) signaling is markedly disturbed in the central nervous system (CNS) of AD patients. Post mortem investigations of brains from patients with AD revealed a markedly downregulated expression of IR, IGF-1R and insulin receptor substrate (IRS) proteins in neurons from AD patients, and these changes progress with severity of neurodegeneration. These findings raise the important question, whether changes in IR/IGF-1R signaling (IIS) are cause, consequence, or may be even compensatory counterregulation to neurodegeneration. In the present thesis the influence of neuronal IGF-1R signaling in the pathophysiology of Alzheimer's disease was analyzed via neuron-specific IGF-1R deletion in an Alzheimer's disease mouse model. Neuron-specific IGF-1R knockout mice were generated using the cre-loxP-system. Mice carrying floxed exon 3 of the IGF-1R gene were crossed with mice expressing the Cre recombinase under control of the neuron-specific synapsin-1 promoter (synCre). Cre-mediated recombination and subsequent excision of exon 3 of the IGF-1 receptor gene results in a frame shift after 213 codons, with an appended sequence of 27 amino acids followed by a stop codon in exon 4. Generated nIGF-1R$^{-/-}$ were further crossed with mice expressing the Swedish mutation of the APP gene (Tg2576). Thus, generated nIGF-1R$^{-/-}$Tg2576 mice were analysed in comparison to wild type, Tg2576 and nIGF-1R$^{-/-}$ animals.

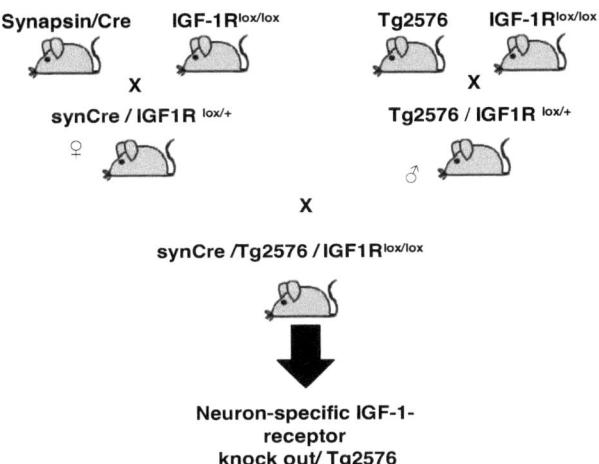

Fig. 3-1 Illustration of the breeding strategy
SynCre and IGF-1R$^{lox/lox}$ mice were crossed to obtain synCre/IGF-1R$^{lox/+}$ females. These females were further crossed with male offspring of mated Tg2576 and IGF-1R$^{lox/+}$ mice to receive neuron-specific IGF-1R knockout in an AD background (SynCre/Tg2576/IGF-1R$^{-/-}$).

3.1 IGF1R expression in cerebellar granule cells of neuron-specific IGF-1R knockout mice (nIGF-1R$^{-/-}$)

The synapsins are peripheral membrane proteins specific for the nervous system. The promoter activity of synapsin-1 was found in different regions of the CNS among these are the hippocampus, the spinal cord and also in the cerebellum[169]. In a first attempt to investigate neuronal deletion efficiency of the IGF-1R knockout, cerebellar granule cells from wild type and nIGF-1R$^{-/-}$ were isolated and maintained in cell culture for 10 days. Afterwards cells were lysed an western blot analyses was performed.

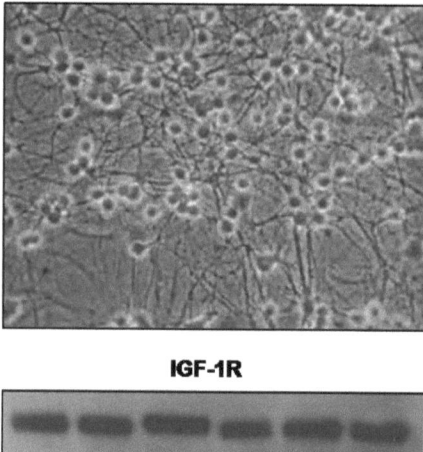

Fig. 3-2 Cerebellar granule cells of nIGF-1R$^{-/-}$ mice
10 days cultured primary cerebellar granule cells. Cells were cultured in basal medium eagle with 10% FCS, 1% P/S, 2mM gluthamine, 25nM KCl and 10μM ara-C. Western blot analysis of IGF-1R in lysed cerebellar granule cells of wild type and nIGF-1R$^{-/-}$ animals. 100μg of protein were applied for western blot analysis.

After isolation and 6 days of selection via cytosine arabinoside (ara-C) a 99% neuronal cell population was achieved (Fig. 3-2). Surprisingly, western blot analysis of 10 days old wildtype and nIGF-1R$^{-/-}$ cerebellar granule cells revealed no differences of IGF-1R expression (Fig 3-2 lower panel). For that reason the pattern of synapsin-1 promoter driven Cre recombinase activity was investigated in detail using a lacZ reporter mouse strain.

3.2 Pattern of Synapsin-1 promoter driven Cre recombinase activity in the CNS

In order to verify neuron-specific and region-specific Cre recombinase expression in the mouse model used in present thesis, X-gal staining was performed in a lac-Z reporter mouse strain. Crossing synCre mice with mice carring the lac-Z reporter gene under the control of the ubiquitously expressed Rosa 26 promoter that is suppressed by a loxP flanked hygromycin resistence gene, which includes a stop cassette, allowed to visualize the pattern of the synapsin-1 Cre recombinase activity and, in consequence, detecting the region were the IGF-1R deletion should occur. Brains of the reporter mice were dissected and prepared for cryo-section and X-gal staining (see material and methods). Afterwards 10-12 μm slices were analyzed using light microscopy.

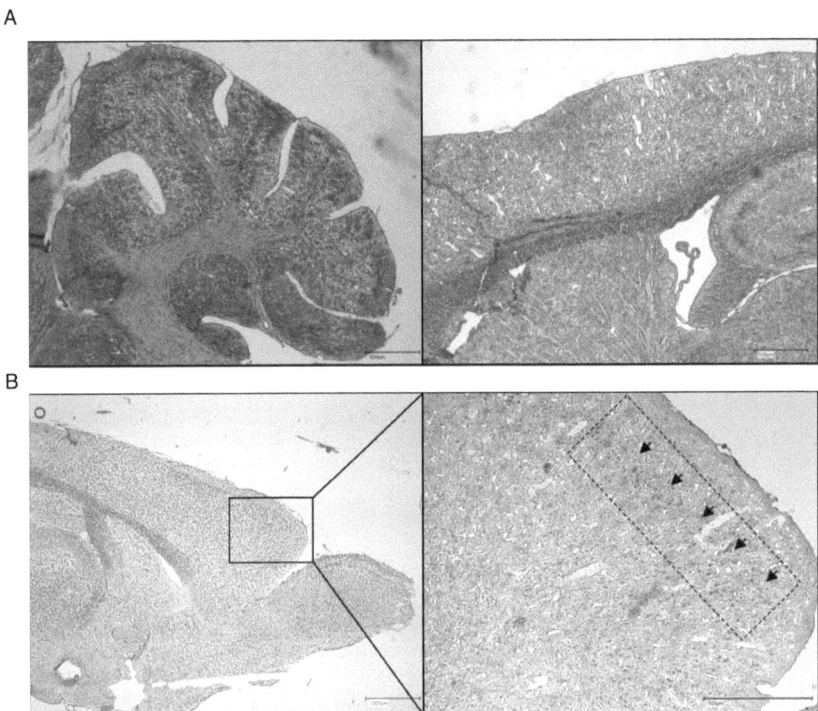

Fig. 3-3 β-Galactosidase staining representing Cre recombinase activity in synCre lacZ reporter mice
A: 25x magnification of β-Galactosidase staining of synapsin cre activity in cerebellum and parietal cortex. B: 25x and 50x magnification of the frontal cortex of synCre lacZ mice. Blue staining indicates the β-galactosidase activity mediated by the synapsin-1 cre promoter in the frontal cortex highlighted by black arrows.

X-gal staining evokes from cleavage of X-Gal by β-galactosidase yielding galactose and 5-bromo-4-chloro-3-hydroxyindole. The latter is oxidized into 5,5'-dibromo-4,4'-dichloro-indigo, an insoluble blue product and can be detected by light microscopy. As shown in Figure 3-3 A, no X-gal staining was found in the cerebellum, parietal cortex, occipital cortex, hypothalamus and olfactory bulb, whereas a very low expression was found in the frontal cortex of the reporter mice, seen in Figure 3-3 B, highlighted by the black arrows. Cre recombinase activity was found predominantly in dentate gyrus (GD) and CA3 region of the hippocampus as shown in different magnifications in Figure 3-4. The blue X-gal staining in the CA3 region and dentate gyrus of hippocampus is well distinguishable from the circumjacent tissue and again highlighted in the lower magnifications by black arrows.

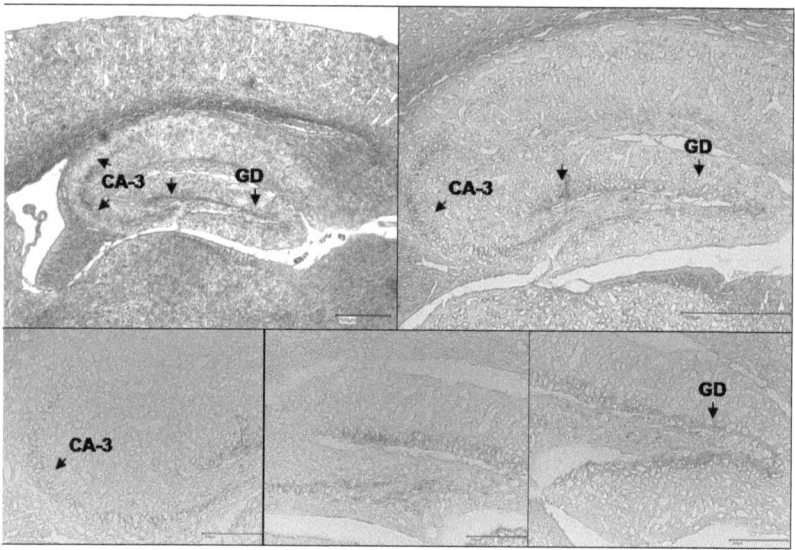

Fig. 3-4 β-galactosidase staining representing Cre recombinase activity in the hippocampal formation of synCre lacZ reporter mice
Upper panels: 25x and 50x magnification of β-Galactosidase staining in hippocampal formation. Lower panels: 100x magnification of hippocampal formation. Blue staining indicates β-galactosidase activity mediated by synapsin-1 promoter in the hippocampus (black arrows). CA-3: cornu ammonis; GD: dentate gyrus

3.3 IGF-1R expression in the CNS and peripheral tissues of nIGF-1R$^{-/-}$ mice

The X-gal staining reveals the dentate gyrus, CA3 region of the hippocampus as major localisation of Cre recombinase expression of SynCre mice. The piriform cortex, frontal cortex and thalamus show Cre recombinase activity to very low extend. In neuron-specific IGF-1R knockout mice IGF-1R deletion was confirmed on protein level by western blot analysis of different brain regions. In order to exclude significant IGF-1R deletion in peripheral tissues western blot analyses from different organs were performed.

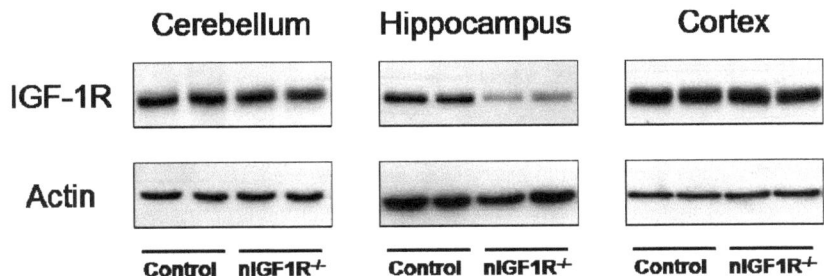

Fig. 3-5 Western blot analysis of IGF-1R protein expression in different brain regions
Western blot analysis of IGF-1R and actin (loading control) expression in lysates of cerebellum, hippocampus and cortex from 60 weeks old wild type and nIGF-1R$^{-/-}$ mice. 100µg of protein were applied on 10% SDS-PAGE gel. Examples of 3 independent experiments.

According to the X-gal staining, as seen in Figure 3-4, a visual IGF-1R deletion occurs only in the hippocampal region. However, caused by IGF-1R expression in non-neuronal cells and remaining IGF-1R expression in the CA1 and CA2 region no complete deletion of IGF-1R was detected from brain lysates of total hippocampus. Other regions like the cerebellum or the cortex show unaltered expression of the IGF-1R protein. The densitometric quantification of the IGF1-1R protein expression in the hippocampus formation seen on the left side of Figure 3-6 indicates a diminished IGF-1R expression to 57% compared to the wild type control group. On the other hand expression of IGF-1R in the cortex was unchanged (seen on the right panel).

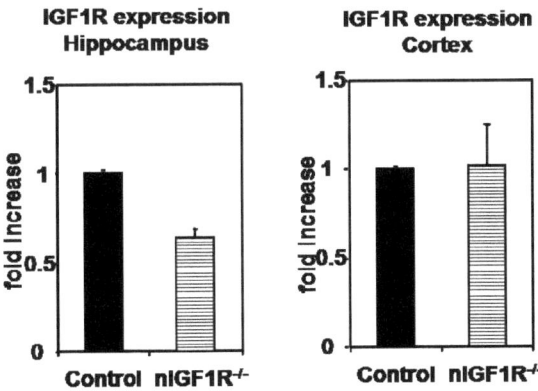

Fig. 3-6 Densitometric quantification of IGF-1R expression in the CNS
Densitometric analysis of IGF-1R expression in hippocampus and cortex of wild type (black bars) and nIGF-1R$^{-/-}$ (black and white striped bars) Data represented mean ± SD (n=8).

Western blot analysis of peripheral tissues shown in Figure 3-7 revealed an unaltered IGF-1R protein expression in heart, lung, kidney, spleen, pancreas, muscle and fat. The IGF-1R protein is known to be not expressed in the liver, so no IGF-1R protein signal could be detected.

In order to investigate the influences of the neuron-specific IGF-1R deletion predominantly in hippocampus the unaffected cortex were used as an additional internal control.

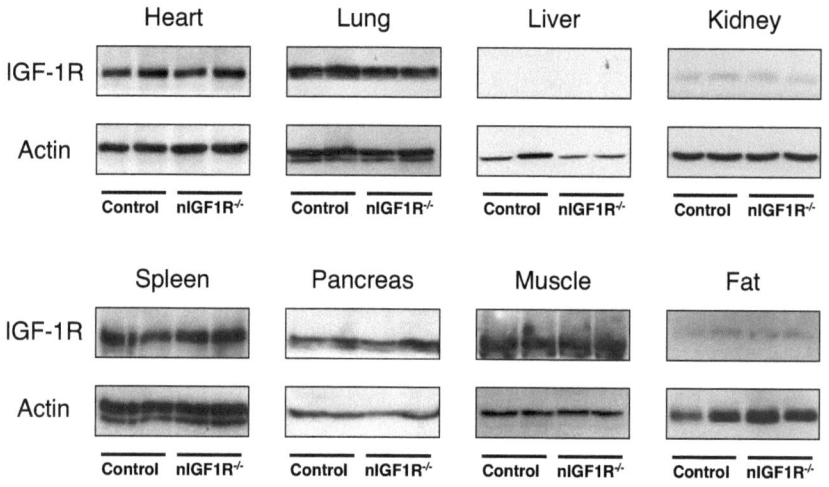

Fig. 3-7 Western blot analysis of IGF-1R protein expression in peripheral tissues
Western blot analysis of IGF-1R and actin (loading control) protein expression in lysates of heart, lung, liver, kidney, spleen, pancreas, muscle and fat from wild type and nIGF-1R$^{-/-}$ mice. 100µg of protein were applied on 10% SDS-PAGE gel. Examples of 3 independent experiments

3.4 IGF-1R signaling in Hippocampus after acute IGF-1 stimulation

To analyse the downstream signals of the IGF-1R hippocampi were stimulated using 10nM IGF-1. Brains of 28 weeks old WT and nIGF-1R$^{-/-}$ mice were dissected and divided sagitally. The hippocampi were dissected and incubated for 10 minutes at

37°C, 5% CO_2 in basal medium eagle with and without 10nM IGF-1. Afterwards Western Blot analyses were performed.

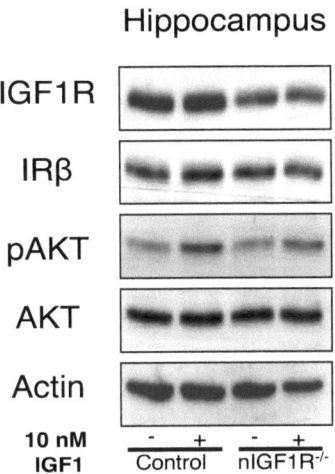

Fig. 3-8 Western blot analysis of IGF-1R expression of Hippocampus and Cortex
Western blot analysis of IGF-1R, insulin receptor, pospho-AKT (Ser 473), AKT and actin protein expression in hippocampus lysates from wild type and nIGF-1R$^{-/-}$ mice. Hippocampi were incubated in Basal medium eagle with and without 10nM IGF for 10 minutes at 37°C, 5% CO_2 and lysed. 100µg of proteinlysates of hippocampi were applied on 10% SDS-PAGE gel. Examples of 3 independent experiments are shown

In contrast to the IGF-1R, IR protein expression was undistinguishable in hippocampi of controls and nIGF-1R$^{-/-}$ mice (Figure 3-8). To simulate a signaling event isolated hippocampi were incubated with and without 10nM IGF. Western blot analysis indicates that hippocampal IGF-1R deletion leads to a decreased IGF-1 stimulated AKT phosphorylation, suggesting that indeed the IGF-1R deletion in nIGF-1R$^{-/-}$ mice reduce downstream signaling.

3.5 Kaplan-Meier analysis

As a result of the Alzheimer's disease, patients as well as animal models have a reduced expectation of life. On the other hand the group of Holzenberger could show that a heterozygous IGF1-R deletion in the whole brain leads to an increase of life-span[170]. To analyse the role of nIGF-1R deletion on mortality in the Tg2576 mice Kaplan-Meier analyses were established.

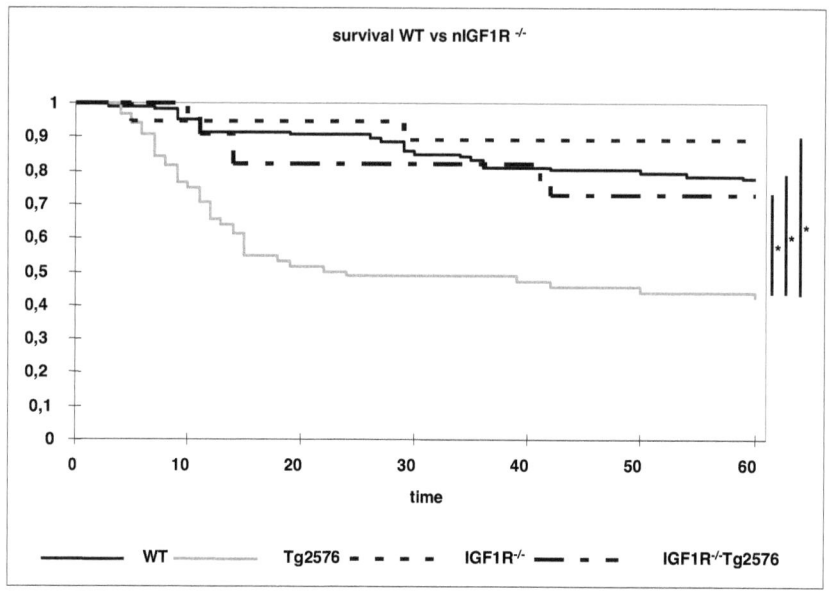

Fig. 3-9 Kaplan-Meier analysis of WT, Tg2576, nIGF-1R$^{-/-}$ and nIGF-1R$^{-/-}$Tg2576 animals
Kaplan-Meier analysis of of WT (n = 108), Tg2576 (n = 64), nIGF-1R$^{-/-}$ (n = 18) and nIGF-1R$^{-/-}$Tg2576 (n = 22) animals. * Wilcoxon rank test p-value ≤ 0,01 versus Tg2576.

Figure 3-9 presents the Kaplan-Meier-analysis of WT, Tg2576, nIGF-1R$^{-/-}$ and nIGF-1R$^{-/-}$ Tg2576 population. Strikingly, nIGF-1R$^{-/-}$ mice were protected against APP^{sw}-induced lethality in Tg2576 background. After 60 weeks of observation approximately 20% of WT and nIGF-1R$^{-/-}$Tg2576 animals died so that no significant differences were observed between WT and nIGF-1R$^{-/-}$Tg2576 nor nIGF-1R$^{-/-}$ animals. In contrast to these nearly 60% of the Tg2576 animals died within 60 weeks, which represents a significant reduced lifespan in comparison to all other genotypes (Wilcox-rank: p-value ≤ 0,01). Remarkably 50% of all Tg2576 animals died already within the first 28 weeks. To further elucidate differences between genders Kaplan-Meier analysis of females and males were done separately (Figure 3-10). In both genders IGF-1R deletion protects Tg2576 mice from premature death. The female population exhibits a slightly higher survival rate than the male population.

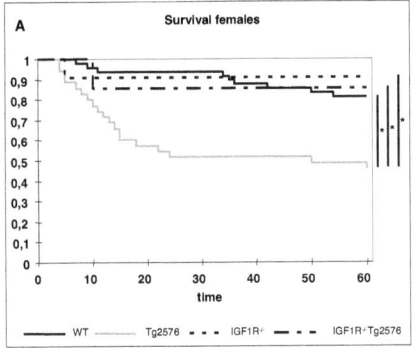 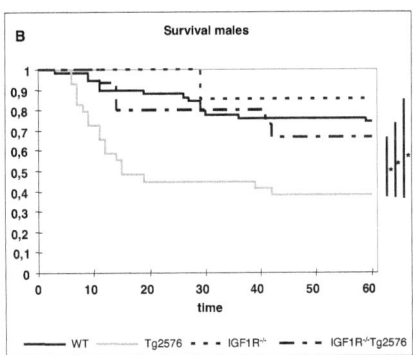

Fig. 3-10 Kaplan-Meier analysis of WT, Tg2576, nIGF-1R$^{-/-}$ and nIGF-1R$^{-/-}$Tg2576 females and males

Left panel: Kaplan-Meier analysis of of WT (n = 48), Tg2576 (n = 35), nIGF-1R$^{-/-}$ (n = 11) and nIGF-1R$^{-/-}$Tg2576 (n = 7) female mice. * Wilcoxon p-value ≤ 0,02 versus Tg2576. Right panel: Kaplan-Meier analysis of of WT (n = 58), Tg2576 (n = 29), nIGF-1R$^{-/-}$ (n = 7) and nIGF-1R$^{-/-}$ Tg2576 (n = 15) male mice. * Wilcoxon rank test p-value ≤ 0,03 versus Tg2576

Concerning the results of Holzenberger group[170], showing a lifespan extension in mice which are hetereozygotus for IGF-1R in all neurons and glia cells of the CNS, hetereozygotus nIGF-1R knockout (nIGF-1R$^{+/-}$) animals were analysed under the hypothesis that nIGF-1R$^{+/-}$ is sufficient to rescue the APP^{sw} -induced lethality. The Kaplan-Meier analysis of Figure 3-11 indicate that after 60 weeks of observation nIGF-1R$^{+/-}$ heterozygosity does not rescue lethality of Tg2576 animals. No difference neither between WT and nIGF-1R$^{+/-}$ nor between Tg2576 and nIGF-1R$^{+/-}$Tg2576 animals was observed. However, there might be a slight shift of the Kaplan-Meier curves, indicating a slight influence of IGF-1R heterozygosity on APP^{sw} induced mortality. Even seperat Kaplan-Meier-Analysis of 17 weeks old animals did not reveal significant differences between the genotypes. Therefore the further investigations were focused on nIGF-1R$^{-/-}$ and nIGF-1R$^{-/-}$Tg2576 animals respectively.

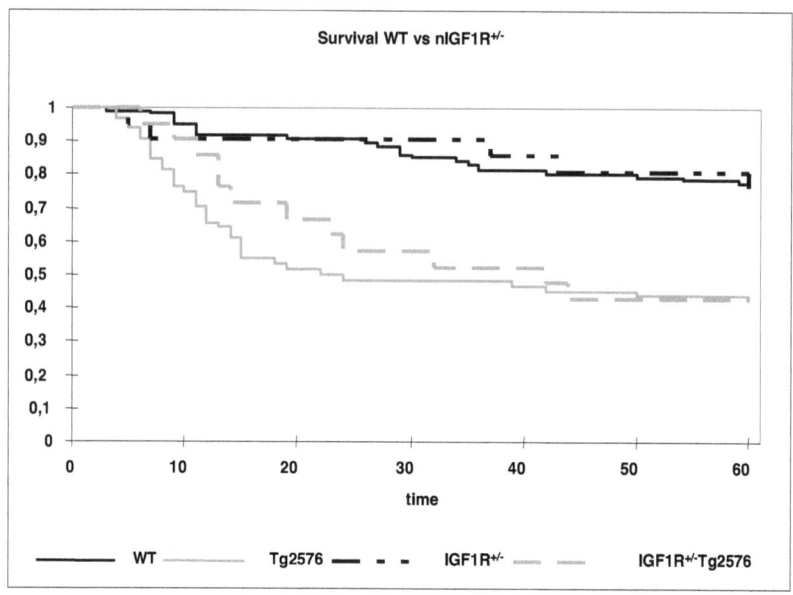

Fig. 3-11 Kaplan-Meier analysis of WT, Tg2576, nIGF-1R$^{+/-}$ and nIGF-1R$^{+/-}$Tg2576 animals

Kaplan-Meier analysis of of WT (n = 108), Tg2576 (n = 64), nIGF-1R$^{+/-}$ (n = 21) and nIGF-1R$^{+/-}$Tg2576 (n = 21) animals.

3.6 Metabolic and somatic characterisation

Glucose metabolism and somatic growth might have influence on life-span and survival. Furthermore, altered glucose metabolism or body growth possibly have influence on the development or progression of AD. Recent studies of Holzenberger group revealed that an IGF-1R$^{-/-}$ deletion in all neurons and all glia cells of the CNS leads to microcephalon, severe growth retardation, infertility, and abnormal behaviour[170]. Furthermore it has been described that whole body deletion of the different IRS proteins, a downstream target of IGF-1R, leads to different pheonotypes. Male mice with a whole body deletion of IRS-2 develop a hyperglycemia and type-2 diabetes and died within 40 weeks[171]. Otherwise IRS-1 deficient mice display growth retardation without developing diabetes. Therefore, glucose metabolism and somatic growth were monitored from week 4 after birth up to week 60.

3.6.1 Glucose homeostasis

In order to assess the influence not only of hippocampal neuron-specific IGF-1R deletion but also of APP^{SW} expression on peripheral glucose homeostasis, blood glucose levels of female and male mice were monitored separately (Figure 3-12). During the observation, up to 60 weeks, no significant alteration of blood glucose levels neither in female nor in male mice of each group was detected.

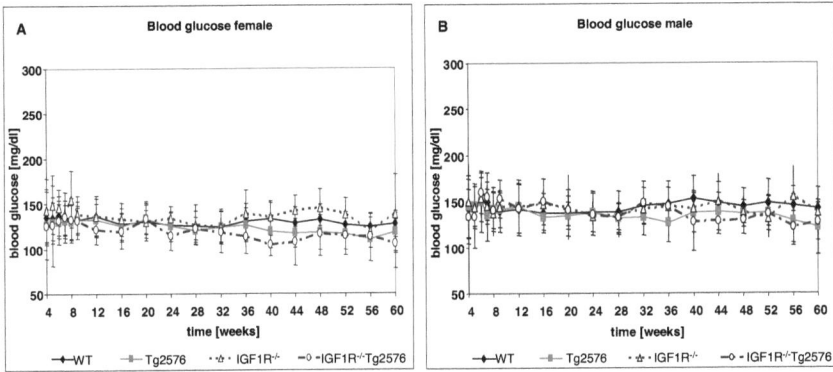

Fig. 3-12 Blood glucose levels of male and female mice during 60 weeks of observation
Average blood glucose levels during observation period in [mg/dl] of WT (black diamond) females (at least n = 36), males (at least n = 42); Tg2576 (grey square) females (at least n = 15), males (at least n = 10); nIGF-1R$^{-/-}$ (white triangle) females (at least n =4) , males (at least n = 4) and nIGF-1R$^{-/-}$Tg2576 (white circle), females (at least n =4), males (at least n = 9). Values are means ± SD

For further evaluation of glucose homeostasis glucose tolerance test were performed at 10 weeks of age. After administration of 2 g glucose/ kg body weight into the peritoneal cavity no significant changes could be detected. All mice of each group displayed a similar increase in blood glucose levels as well as a similar clearance after glucose challenge, so that all animals returned to normal blood glucose levels after 120 min (see Figure 3-13).

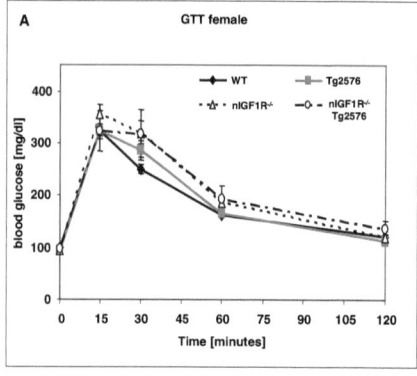 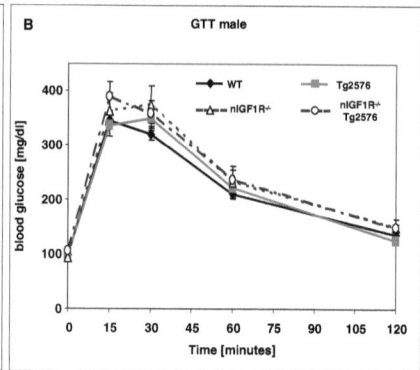

Fig. 3-13 Glucose tolerance test of male and female mice
Average blood glucose levels in [mg/dl] of 10 weeks old WT (black diamond) females (n = 65), males (n = 69); Tg2576 (grey square) females (n = 36), males (n = 26); nIGF-1R$^{-/-}$ (white triangle) females (n = 16), males (n = 11) and nIGF-1R$^{-/-}$Tg2576 (white circle) females (n = 10), males (n = 17). Values are means ± SD

In addition, Insulin sensitivity was determined by insulin tolerance tests at 11-12 weeks of age. Animals of each group received 0.75 U/kg body weight of human insulin into the peritoneal cavity. As a result of the insulin administration blood glucose levels decreased within 60 minutes after administration (cp. Figure 3-14). No significant changes were observed.

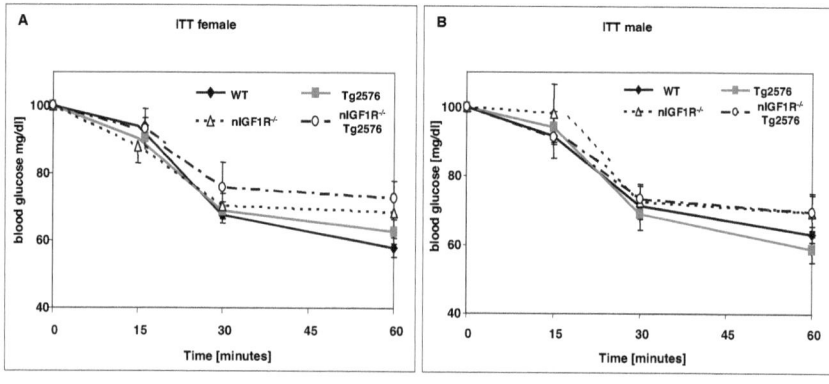

Fig. 3-14 Insulin tolerance test from male and female mice
Average blood glucose levels in [mg/dl] of 11 weeks old WT (black diamond) females (n = 60), males (n = 64); Tg2576 (grey square) females (n = 37), males (n = 22); nIGF-1R$^{-/-}$ (white triangle) females (n = 16), males (n = 10) and nIGF-1R$^{-/-}$Tg2576 (white circle) females (n = 9), males (n = 15). Values are means ± SD

The results of the metabolic characterisation lead to the conclusion that neither the IGF-1R deletion nor Tg2576 background has an impact on peripheral glucose homeostasis.

3.6.2 Somatic characterisation

Kappeler et al. showed that deletion of IGF-1R in all neuron and all glia cells leads to microcephalon and severe growth retardation[170]. Heterozygosity of IGF-1R in neurons and glia cells caused reduced adult body size, metabolic alterations and led to delayed mortality and a longer mean lifespan. Whole body deletion of the IRS-1 gene causes growth retardation as well and like other dwarf mice IRS-1$^{-/-}$ mice show a life-span extension compared to their wild type littermates. Hence, body size, brain weight and brain-body ratio of adult 60 weeks old mice were measured and compared to their littermates. In addition, body weights of each group were monitored during the observation period of 60 weeks.

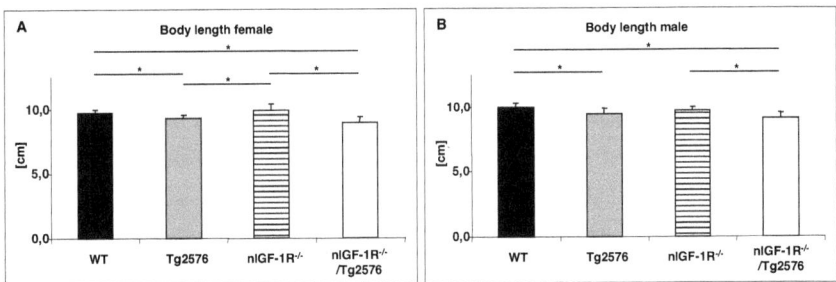

Fig. 3-15 Body length of females and males of the study group
Average body size of 60 weeks old WT (black bars), females (n = 29), males (n = 31); Tg2576 (grey bars) females (n = 6), males (n = 12); nIGF-1R$^{-/-}$ (white striped bars) females (n = 6), males (n = 6) and nIGF-1R$^{-/-}$Tg2576 (white bars) females (n = 6), males (n = 4). Values are means ± SD, * unpaired Student's t-test p-value ≤ 0,02).

Figure 3-15 presents the body length of female and male mice of each group. Female mice in a Tg2576 background with and without nIGF-1R deletion have a significanttly decreased body length. Tg2576 female mice are approximately 5% smaller than WT or nIGF-1R$^{-/-}$ animals. Compared to nIGF-1R$^{-/-}$Tg2576 animals Tg2576 females are 4% taller. Furthermore nIGF-1R$^{-/-}$Tg2576 females are 8-10% smaller compared to WT or nIGF-1R$^{-/-}$ females. No significant changes were detected between WT and nIGF-1R$^{-/-}$ females. In male mice significant changes of body length were detected between WT and Tg2576 as well as between WT and nIGF-1R$^{-/-}$ Tg2576. Tg2576 males were 5% smaller and nIGF-1R$^{-/-}$Tg2576 9% smaller then WT males. In addition nIGF-1R$^{-/-}$ males were 6% taller than nIGF-1R$^{-/-}$Tg2576. Like female mice no significant changes were detected between WT and nIGF-1R$^{-/-}$.

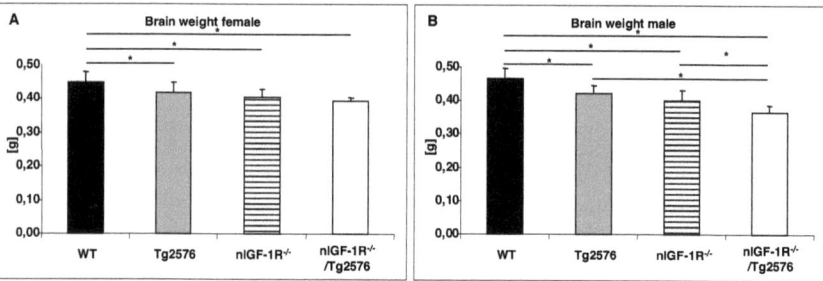

Fig. 3-16 Brain weight of females and males of the study group
Average brain weight of 60 weeks old WT (black bars), females (n = 29), males (n = 31); Tg2576 (grey bars) females (n = 6), males (n = 12); nIGF-1R$^{-/-}$ (white striped bars) females (n = 6), males (n = 6) and nIGF-1R$^{-/-}$Tg2576 (white bars) females (n = 6), males (n = 4). Values are means ± SD, * unpaired Student's t-test p-value ≤ 0,008).

Among mammals including mice, the size of the brain and its components are related to body size[172]. For 230 different species of mice, the brain-body ratio is constant at the same age[173]. At least 50% of brain and body growth is mediated by the insulin-IGF-signaling system[174]. Thus, brain weight of 60 weeks old mice was measured and plotted in Figure 3-16. The average brain weight of Tg2576 females was about 7% decreased and the brain weight of nIGF-1R$^{-/-}$Tg2576 females even about 12% compared to wild type mice. Surprisingly, the brains weight of nIGF-1R$^{-/-}$ were about 12% decreased compared to WT females as well. Furthermore brain weight of males Tg2576, nIGF-1R$^{-/-}$ and nIGF-1R$^{-/-}$Tg2576 were significant decreased compared to WT. In detail average brain weight of Tg2576 males was about 10% reduced and brain weight of nIGF-1R$^{-/-}$ about 14% reduced. Strikingly the average brain weight of nIGF-1R$^{-/-}$Tg2576 was about 9%, 13% and even about 21% reduced compared to nIGF-1R$^{-/-}$, Tg2576 and WT respectively.

The alterations of body size were accompanied by changes in body weight. In the upper panel of Figure 3-17 (A-B) body weights of 60 weeks old female and male mice are shown.

Compared to body length the reductions in body weight are unproportional higher in Tg2576 and nIGF-1R$^{-/-}$Tg2576 animals as in wild type and nIGF-1R$^{-/-}$ animals. For example Tg2576 female mice displayed a 5% reduced body size but the body weight is even about 19% reduced. Similar nIGF-1R$^{-/-}$Tg2576 females body size was about 8% reduced but the body weight about 18%. A similar result was observed by comparing nIGF-1R$^{-/-}$ with Tg2576 and nIGF-1R$^{-/-}$Tg2576 females. According to the results in female similar changes in body weight was observed in male mice but the results were more pronounced. Compared with WT males the Tg2576 males reveal a 5% body size reduction but a 22% reduced body weight. The nIGF-1R$^{-/-}$Tg2576 male mice exhibit 9% decreased body size and 28% decreased body weight. However, no differences were detected between Tg2576 and nIGF-1R$^{-/-}$Tg2576

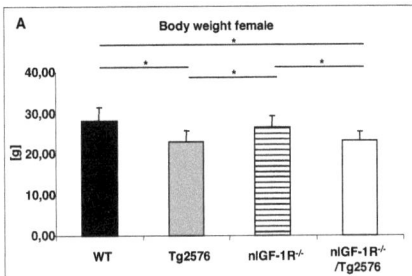
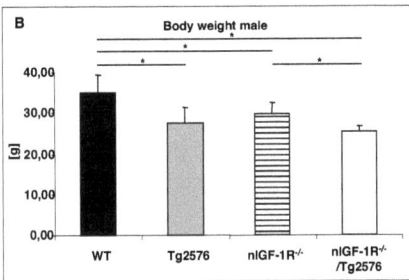
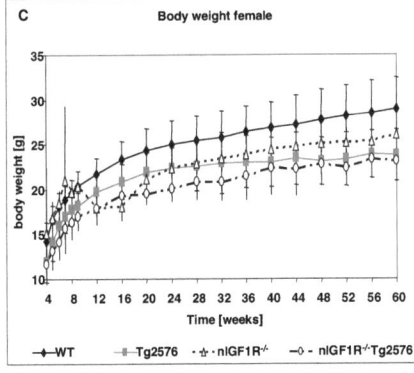
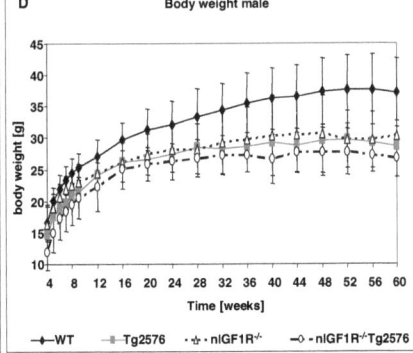

Fig. 3-17 Body weight of 60 weeks old animals and growth curves of the different genotypes

Upper panel: Average body weight of WT of 60 weeks old (black bars), females (n = 29), males (n = 31); Tg2576 (grey bars) females (n = 6), males (n = 12); nIGF-1R$^{-/-}$ (white striped bars) females (n = 6), males (n = 6) and nIGF-1R$^{-/-}$Tg2576 (white bars) females (n = 6), males (n = 4). Values are means ± SD, * unpaired Student's t-test p-value ≤ 0,04). Lower panel: Average blood glucose levels during observation period of WT (black diamond) females (at least n = 36), males (at least n = 42); Tg2576 (grey square) females (at least n = 15), males (at least n = 10); nIGF-1R$^{-/-}$ (white triangle) females (at least n =4) , males (at least n = 4) and nIGF-1R$^{-/-}$Tg2576 (white circle), females (at least n =4), males (at least n = 9). Values are means ± SD

Growth curves of the different genotypes starting at the age of 4 weeks up to 60 weeks displayed a similar trend as seen in body weight evaluation of 60 weeks old animals (Figure 3-17 C-D). Significant changes in body weight are according to this no single event but rather a result of the different genetic modifications. Additionally this effect seems to be more pronounced in male than female mice.

Due to the significant changes in body weight, body composition was examined by nuclear magnetic resonance analysis (NMR) at the age of 28 and 60 weeks.

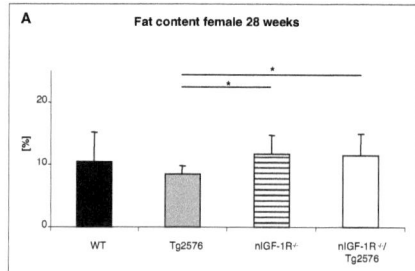
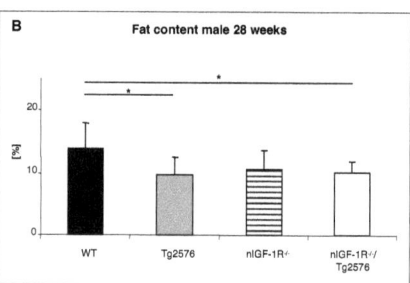

Fig. 3-18 Fat content at 28 weeks
Average fat content of 28 weeks old WT (black bars), females (n = 27), males (n = 17); Tg2576 (grey bars) females (n = 13), males (n = 4); nIGF-1R$^{-/-}$ (white striped bars) females (n = 9), males (n = 4) and nIGF-1R$^{-/-}$Tg2576 (white bars) females (n = 9), males (n = 10). Values are means ± SD, * unpaired Student's t-test p-value ≤ 0,01.

Compared to the results of body weight in female mice a significant difference in fat content between Tg2576 and nIGF-1R$^{-/-}$ and nIGF-1R$^{-/-}$Tg2576 animals was noticed. nIGF-1R$^{-/-}$ females displayed ~30% more fat as Tg2576 females. In contrast to body weight no fat content reduction were observed in nIGF-1R$^{-/-}$Tg2576 females compared to wild type and nIGF-1R$^{-/-}$ females. Thus nIGF-1R$^{-/-}$Tg2576 females displayed a ~30% higher fat content compared to Tg2576 females. In male mice significant changes were only detectable in Tg2576 and nIGF-1R$^{-/-}$Tg2576 animals comparted to WT. In males the fat contents of Tg2576 as well as nIGF-1R$^{-/-}$Tg2576 were about 30% reduced compared to wild type and nIGF-1R$^{-/-}$ male mice.

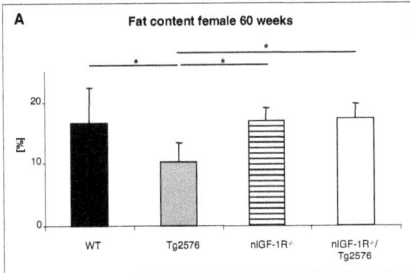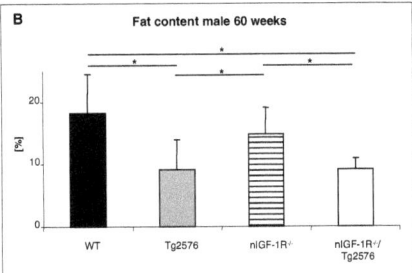

Fig. 3-19 Fat content at 60 weeks
Average fat content of 60 weeks old WT (black bars), females (n = 27), males (n = 36); Tg2576 (grey bars) females (n = 15), males (n = 6); nIGF-1R$^{-/-}$ (white striped bars) females (n = 6), males (n = 6) and nIGF-1R$^{-/-}$Tg2576 (white bars) females (n = 4), males (n = 8). Values are means ± SD, * unpaired Student's t-test p-value ≤ 0,004.

During aging an increase in fat proportion is physiological, as seen in the evaluation of 60 weeks old mice and illustrated in Figure 3-19. This increase in fat content was not observed for all genotypes. Tg2576 females and males as well as nIGF-1R$^{-/-}$ Tg2576 males did not increase their fat proportion. Tg2576 females exhibited a 60% of fat mass in comparison to WT, nIGF-1R$^{-/-}$ and nIGF-1R$^{-/-}$Tg2576. In male mice this difference was more pronounced. Here, the Tg2576 animals as well as nIGF-1R$^{-/-}$ Tg2576 exhibited only 50% of fat mass compared to WT animals.
As shown in Figure 3-20 A only Tg2576 females accumulate less fat during aging compared to the other genotypes. More exciting results were observed in 60 weeks old Tg2576 and nIGF-1R$^{-/-}$Tg2576 males (Figure 3.20 B). Tg2576 and nIGF-1R$^{-/-}$ Tg2576 did not increase their fat proportion during aging in contrast to wild type or nIGF-1R$^{-/-}$ mice.
In conclusion the Tg2576 background has an influence on body fat mass during aging which is more pronounced in males as in females. Remarkable in Tg2576 females IGF-1R deletion rescue the lacking fat accumulation during aging. All together it is rather unlikely that this effect influence survival rate of nIGF-1R$^{-/-}$Tg2576 because the survival benefit is present in male and in female.

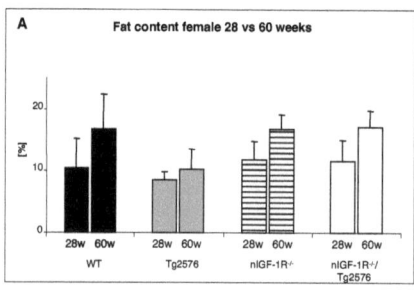 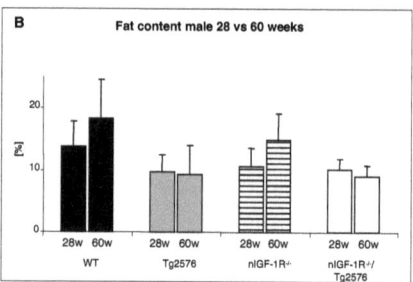

Fig. 3-20 Comparison of Fat content of 28 and 60 weeks old animals
Average fat content of 28 and 60 weeks old WT (black bars); Tg2576 (grey bars); nIGF-1R$^{-/-}$ (white striped bars) and nIGF-1R$^{-/-}$Tg2576 (white bars) females and male mice. Values are means ± SD.

In order to evaluate whether the observed changes in brain weight and body weight alter the proportion of brain tissue in relation to body weight brain-body ratio was gender- and genotype-specific calculated and compared. However, not only aging might influence body weight, but also diet and different other important factors as well. In the present evaluation diet can be excluded because all animals received a normal standard diet.

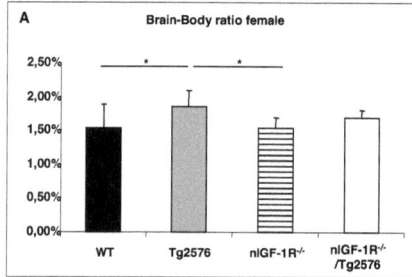 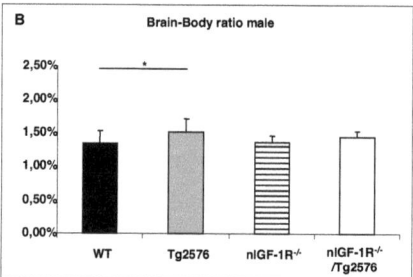

Fig. 3-21 Brain-Body ratio of 60 weeks old mice
Average brain-body ratio of 60 weeks old WT (black bars), females (n = 29), males (n = 31); Tg2576 (grey bars) females (n = 6), males (n = 12); nIGF-1R$^{-/-}$ (white striped bars) females (n = 6), males (n = 6) and nIGF-1R$^{-/-}$Tg2576 (white bars) females (n = 6), males (n = 4). Values are means ± SD, * unpaired Student's t-test p-value ≤ 0,04).

As indicated in Figure 3-21 slight but significant differences were detected between female and male WT and Tg2576 mice as well as between Tg2576 and nIGF-1R$^{-/-}$

females. However, no changes were observed between Tg2576 and nIGF-1R$^{-/-}$ Tg2576 mice.

The somatic and metabolic characterisation of the study groups revealed no differences between Tg2576 and nIGF-1R$^{-/-}$Tg2576 mice, explaining increased survival of nIGF-1R$^{-/-}$Tg2576 animals in comparison to Tg2576 animals.

3.7 Biochemical analysis of 28 weeks old animals

Protein expression analysis of nIGF-1R$^{-/-}$ mice results in downregulation of IGF-1R specificly in the hippocampus. Concomitant with the downregulation of IGF-1R a significant higher survival of nIGF-1R$^{-/-}$Tg2576 was observed in comparison to Tg2576 animals. In order to investigate the influence of downregulated IIS in an early phase of AD development biochemical analyses were performed at the age of 28 weeks.

3.7.1 Analysis of IGF-1R/IR signaling

First the IIS of 28 weeks old animals was investigated. The expressions of key components of the IIS were analysed.

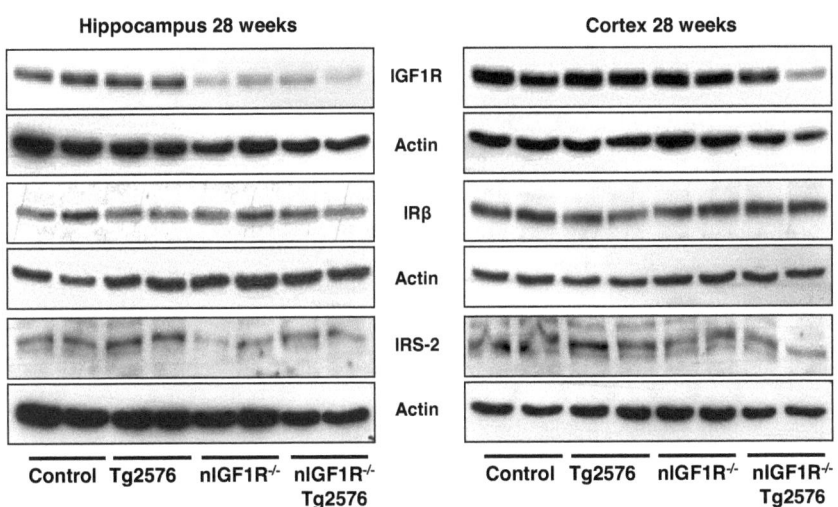

Fig. 3-22 Western blot analysis of IGF1-R/IR signaling of 28 weeks old mice I
Western blot analysis of IGF-1R, IR, IRS-2 and actin (loading control) protein expression in Hippocampus and cortex lysates from 28 weeks old wild type, Tg2576, nIGF-1R$^{-/-}$ and nIGF-

1R⁻/⁻Tg2576 mice. 100μg of protein were applied on 10% SDS-PAGE. Examples of 2 independent experiments.

As shown in Figures 3-5 and 3-6 IGF-1R was downregulated in the hippocampus but not in the cortex of nIGF-1R$^{-/-}$ and nIGF-1R$^{-/-}$Tg2576 animals. The expression of the IR was not affected but a slight reduction of the downstream target protein IRS-2 was observed (cp figure 3-22). Consequentially the further downstream proteins were analysed but no changes were observed in AKT and in the steady state level of Ser^{473} phosphorylated AKT. Investigations of the downstream kinase GSK-3β, which might be involved in the regulation of secretases, reveals also no changes at the age of 28 weeks. Also no changes were detected in the protein expression of the extracellular signal regulated kinase ERK-1/2. Furthermore, ERK-1/2 phosphorylation was slight higher in Tg2576 and nIGF-1R$^{-/-}$ Tg2576 compared to WT and nIGF-1R$^{-/-}$ mice.

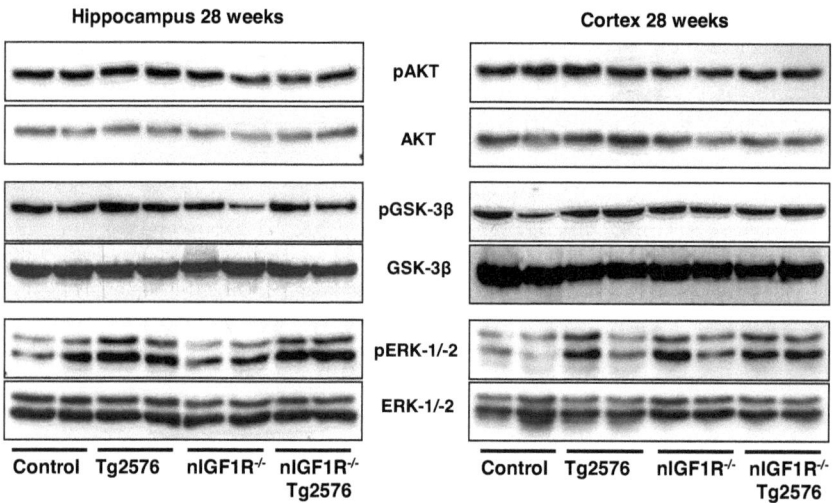

Fig. 3-23 Western blot analysis of IGF1-R/IR signaling of 28 weeks old mice II
Western blot analysis of pospho-AKT (Ser^{473}), AKT (loading control), pospho-GSK-3β (Ser^9), GSK-3β (loading control), pospho-ERK-1/2 (Thr^{202} / Tyr^{204}), ERK-1/2 (loading control) protein expression in Hippocampus and cortex lysates from 28 weeks old wild type, Tg2576, nIGF-1R$^{-/-}$ and nIGF-1R$^{-/-}$Tg2576 mice. 100μg of protein were applied on 10% SDS-PAGE. Examples of 2 independent experiments.

3.7.2 Investigation of APP processing

Subsequently APP processing was analysed. APP cleavage is responsible for Aβ peptide generation, which in turn is responsible for plaque formation. As seen in Figure 3-23 no changes in the expression of the amyloid precursor protein was detected between the appropriate genotype. However, the occurrence of the C-terminal cleavage products (CTFs) produced by α- and β-secretase cleavage were markedly reduced in nIGF-1R$^{-/-}$Tg2576. As indicated in figure 3-24 (lower panel) the α- and β-CTFs of the nIGF-1R$^{-/-}$Tg2576 animals were about 50% reduced in the hippocampus compared to Tg2576. However, quantification of CTFs in the cortex revealed a minor reduction which failed to reach significance.

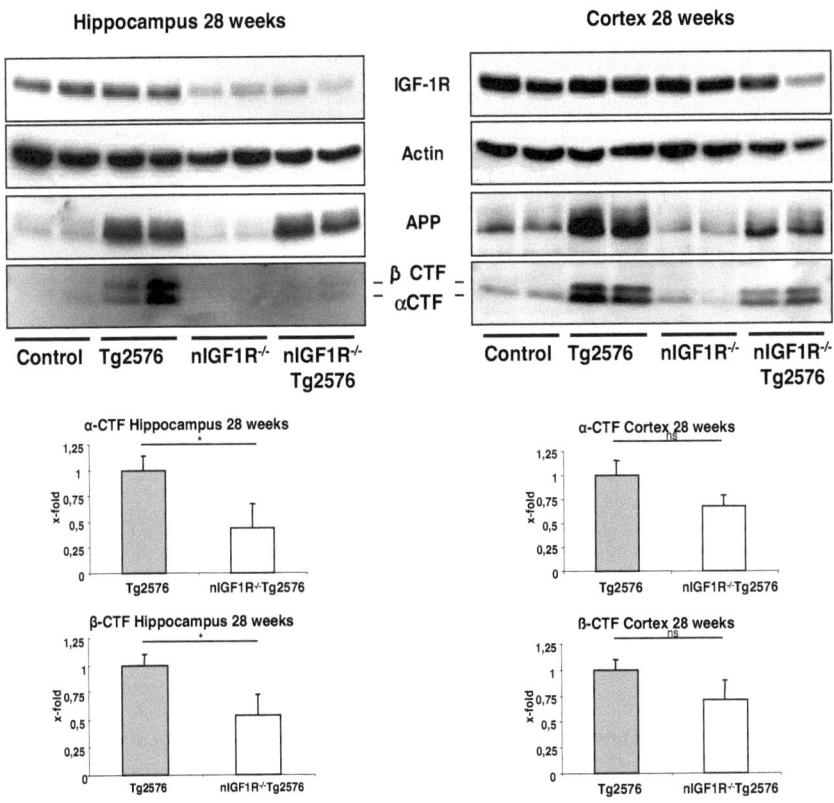

Fig. 3-24 Western blot and densitometric analysis of APP processing of 28 weeks old mice

Upper panel: Western blot analysis of APP, actin (loading control), α- and β-C-terminal fragments appearance in hippocampus and cortex lysates from 28 weeks old wild type, Tg2576, nIGF-1R$^{-/-}$ and nIGF-1R$^{-/-}$Tg2576 mice. 100μg of protein were applied on 10% SDS-PAGE or 15% SDS-PAGE respectively. Examples of 2 independent experiments. Lower panel: densitometric quantification of α- C-terminal fragments and β-C-terminal fragments protein expression from Tg2576 (grey bars) and nIGF-1R$^{-/-}$Tg2576 (white bars) mice. Values are means ± SD, n = 4, unpaired Student's t-test p-value ≤ 0,05.

In this context the small cleavage products were aim of further investigation as indicated in Figure 3-25. Despite of all efforts it was not possible to clearly detect the 4-5 kDa Aβ$_{1-40/42}$ monomer peptides in western blot analysis. However, 15kDa Aβ1-40/42 peptides representing trimeres could be detected as shown in Figure 3-25. Aβ$_{1-40/42}$ trimeres in the hippocampal region of nIGF-1R$^{-/-}$Tg2576 mice displayed a clear reduction in comparison to Tg2576 animals. In the cortex were no noticeable differences. To clarify the Aβ$_{1-40/42}$ peptides content Enzyme Linked Immunosorbent Assays (ELISA) were performed. Figure 3-25 (lower panel) displays the results of the ELISA for the Aβ$_{1-40}$ peptides. A significant reduction of ~50% in the hippocampus of nIGF-1R$^{-/-}$Tg2576 compared to Tg2576 mice was observed. No differences were detectable in the cortex between Tg2576 and nIGF-1R$^{-/-}$Tg2576 animals.

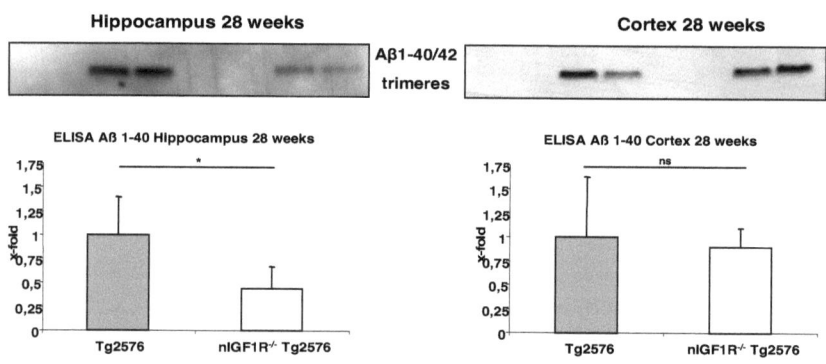

Fig. 3-25 Western blot and ELISA analysis of Amyloid-β in 28 weeks old mice
Upper panel: Western blot analysis of Amyliod-β$_{1-40/42}$ appearance in hippocampus and cortex lysates from 28 weeks old wild type, Tg2576, nIGF-1R$^{-/-}$ and nIGF-1R$^{-/-}$Tg2576 mice. 100μg of protein were applied on 15% SDS-PAGE. Examples of 2 independent experiments. Lower panel: ELISA analysis of Amyliod-β$_{1-40}$ in the hippocampus and cortex from 28 weeks old Tg2576 (grey bars) and nIGF-1R$^{-/-}$Tg2576 (white bars) mice. Values are means ± SD, n = 4, * unpaired Student's t-test p-value ≤ 0,05.

The reduced CTFs and $A\beta_{1-40}$ peptides might caused by changes in expression or activity of proteins involved in synthesis or clearances of these products. To further investigate this point expression levels of Beta-site APP cleaving enzyme-1 (BACE), Insulin degrading enzyme (IDE) and Apolipoprotein E (ApoE) were investigated. The expression level of BACE-1 as putative β-secretase was not altered in the brains of 28 weeks old animals. The expression of IDE and ApoE were also not changed.

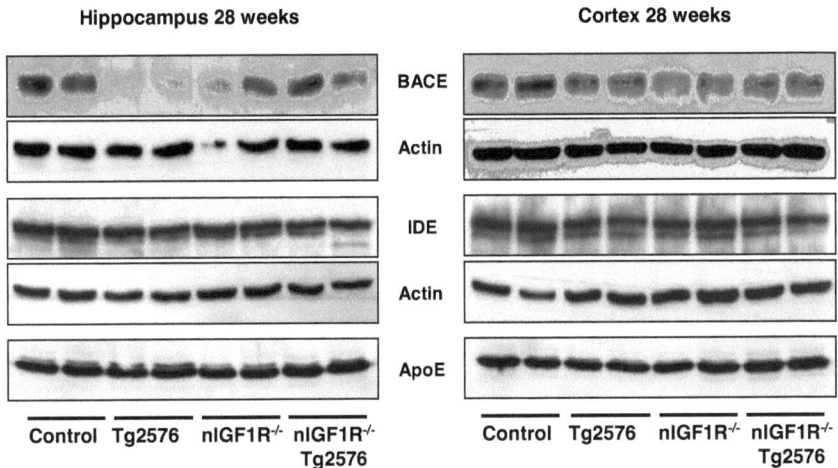

Fig. 3-26 Western blot analysis of proteins involved in APP cleavage and Aβ clearance
Western blot analysis of BACE, IDE, ApoE and actin (loading control) protein expression in Hippocampus and cortex lysates from 28 weeks old wild type, Tg2576, nIGF-1R$^{-/-}$ and nIGF-1R$^{-/-}$Tg2576 mice. 100μg of protein were applied on 10% SDS-PAGE. Examples of 2 independent experiments.

Based on the results of biochemical analysis of 28 weeks old animals a 50% reduced expression of APP cleavage products like CTFs and $A\beta_{1-40}$ were detected without obvious change in expression of proteins involved in IIS, Aβ clearance or cleavage of APP.

3.8 Biochemical analysis of 60 weeks old animals

In order to investigate the influence of IIS on disease progression of AD and APP metabolism in the hippocampus and cortex 60 weeks old animals were analysed.

3.8.1 Analysis of IGF-1R/IR signaling

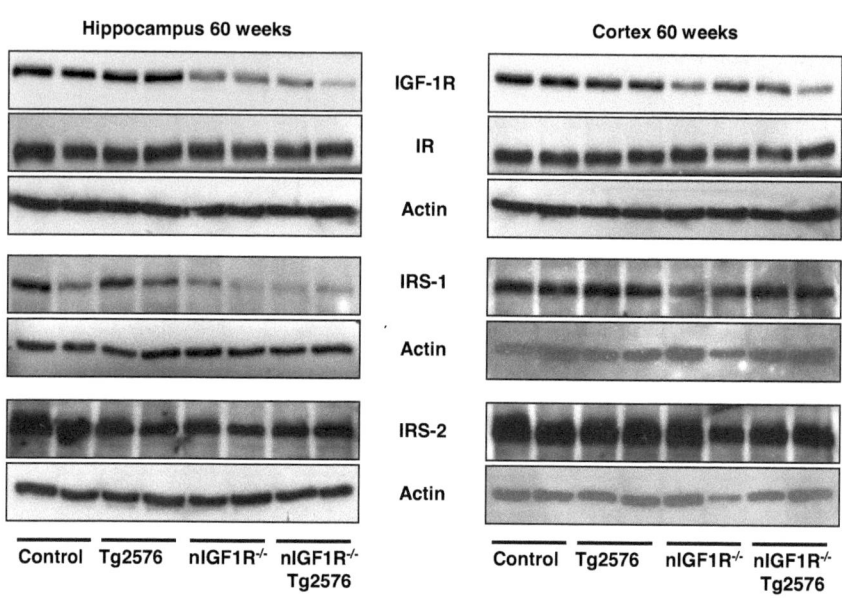

Fig. 3-27 Western blot analysis of IGF1-R/IR signaling of 60 weeks old mice
Western blot analysis of IGF-1R, IR, IRS-1, IRS-2 and actin (loading control) protein expression in Hippocampus and cortex lysates from 60 weeks old wild type, Tg2576, nIGF-1R$^{-/-}$ and nIGF-1R$^{-/-}$Tg2576 mice. 100µg of protein were applied on 10% SDS-PAGE. Examples of 2 independent experiments.

Analyses of brain lysates of 60 weeks old animals revealed similar results as in 28 weeks old animals. IGF-1R expression was reduced in the nIGF-1R$^{-/-}$ and nIGF-1R$^{-/-}$ Tg2576 animals whereas no changes were detected in IR expression. In contrast to 28 weeks old animals no changes in IRS-2 protein expression were seen in 60 weeks old animals. Surprisingly, a downregulation of IRS-1 protein expression was detected in hippocampus of animals with the IGF-1R deletion (nIGF-1R$^{-/-}$ and nIGF-1R$^{-/-}$Tg2576). No downregulation of IRS-1 expression was observed in the cortex. To quantify the amount of protein expression densitometric analyses were performed.

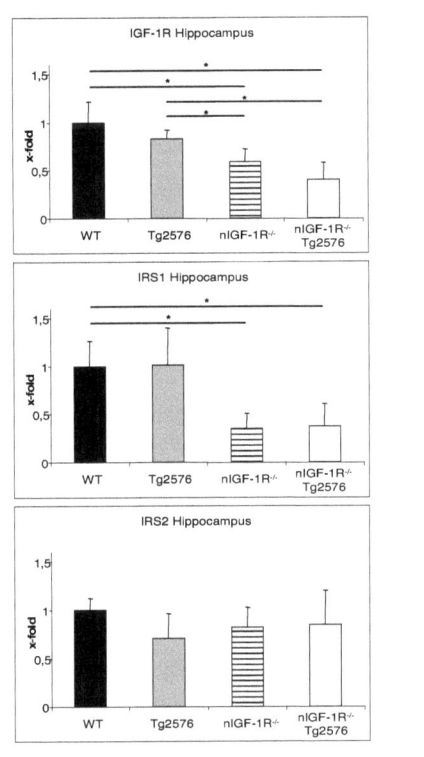
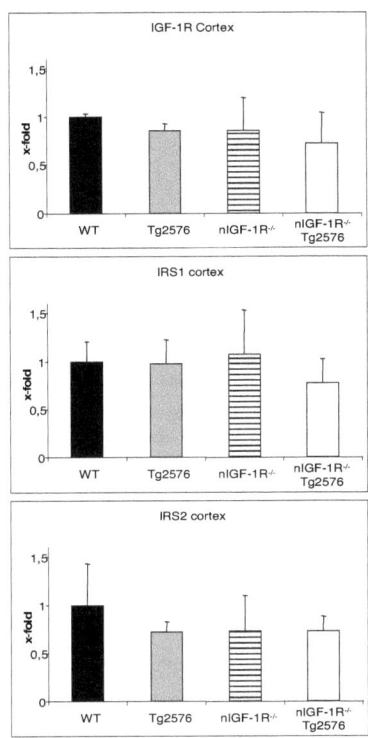

Fig. 3-28 Densitometric quantification of IGF-1R, IRS-1 and IRS-2 protein expression in 60 weeks old mice

Densitometric quantification of IGF-1R, IRS-1 and IRS-2 protein expression from 60 weeks old wild type (black bars), Tg2576 (grey bars), nIGF-1R$^{-/-}$ (white striped bars) and nIGF-1R$^{-/-}$ Tg2576 (white bars) mice. Values are means ± SD, n = 4, * unpaired Student's t-test p-value ≤ 0,05.

As indicated in Figure 3-28 an approximately 50% downregulation of IGF-1R was detected in the hippocampus only. As observed in the WB analysis a 60% reduction of IRS-1 protein expression was detected in the hippocampi of nIGF-1R$^{-/-}$ and nIGF-1R$^{-/-}$Tg2576 animals. No significant changes were quantified in WT and Tg2576 mice as well as in corteces of all animals. IRS-2 expression levels were reduced by 20-30% in Tg2576, nIGF-1R$^{-/-}$ and nIGF-1R$^{-/-}$ mice in comparison to wild type mice. These results could be shown in hippocampus as well as in cortex but failed to reach significant. Thus direct correlation between nIGF-1R$^{-/-}$ and downregulation of IRS-1 was observed. Furthermore the protein kinase ERK-1/2 was analysed.

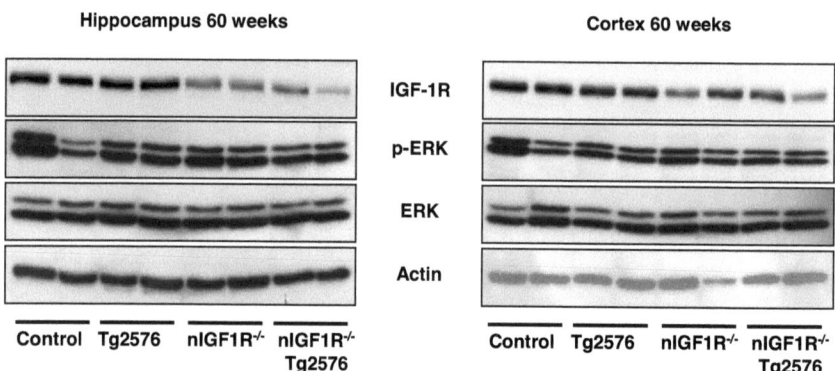

Fig. 3-29 Western blot analysis of ERK-1/2
Western blot analysis of IGF-1R, phospho-ERK-1/2 (Thr^{202} / Tyr^{204}), ERK-1/2 and actin (loading control) protein expression in Hippocampus and cortex lysates from 60 weeks old wild type, Tg2576, nIGF-1R$^{-/-}$ and nIGF-1R$^{-/-}$Tg2576 mice. 100µg of protein were applied on 10% SDS-PAGE. Examples of 2 independent experiments.

Like the results of the 28 weeks old animals no apparent difference was detected in ERK-1/2 expression and phosphorylation. As in 28 weeks old annimals no alteration of the phosphorylated and unphosphorylated AKT proteins were found. Protein expression of the phosphatase PTEN was unchanged. PTEN might regulate the phosphorylation of GSK-3 (Ser^9) in an AKT independent manner.

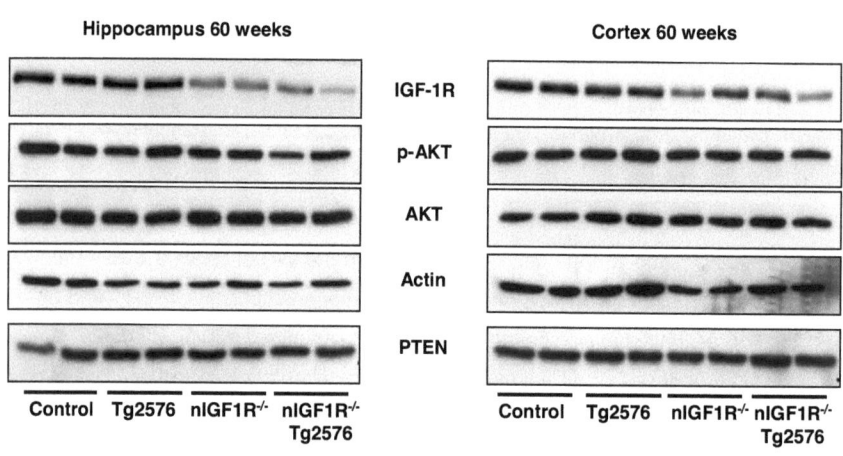

Fig. 3-30 Western blot analysis of AKT and PTEN of 60 weeks old mice
Western blot analysis of IGF-1R, phospho-AKT (Ser^{473}), AKT, PTEN and actin (loading control) protein expression in Hippocampus and cortex lysates from 60 weeks old wild type, Tg2576, nIGF-1R$^{-/-}$ and nIGF-1R$^{-/-}$Tg2576 mice. 100μg of protein were applied on 10% SDS-PAGE. Examples of 2 independent experiments.

Subsequently GSK-3β was analysed. Although GSK-3β protein expression was not altered in the hippocampus, downregulation of the phosphorylated form of GSK-3β as well as GSK-3α were observed.

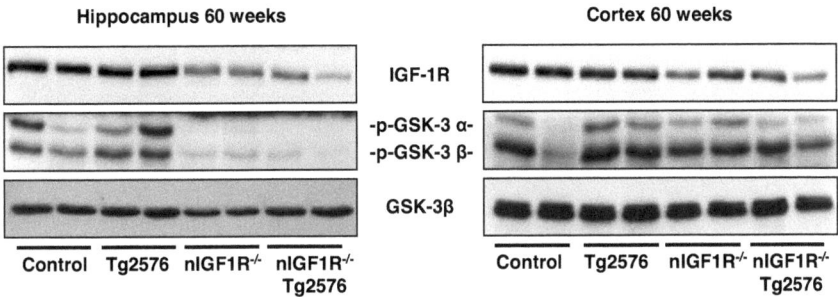

Fig. 3-31 Western blot analysis of GSK-3
Western blot analysis of IGF-1R, phospho-GSK-3α/β (Ser$^{21/9}$), GSK-3β (loading control) protein expression in Hippocampus and cortex lysates from 60 weeks old wild type, Tg2576, nIGF-1R$^{-/-}$ and nIGF-1R$^{-/-}$Tg2576 mice. 100μg of protein were applied on 10% SDS-PAGE. Examples of 2 independent experiments.

Furthermore the expression and phosphorylation of the transcription factor Foxo1 was investigated. Downregulation of Foxo1 orthologs in *Caenorhabditis elegans* leads to lifespan extension and thus might be a candidate for survival rescue of nIGF-1R$^{-/-}$Tg2576 animals. However, WB analysis of Foxo1 and phosphorylated Foxo1 could not reveal any changes in the hippocampus and cortex of the investigated animals.

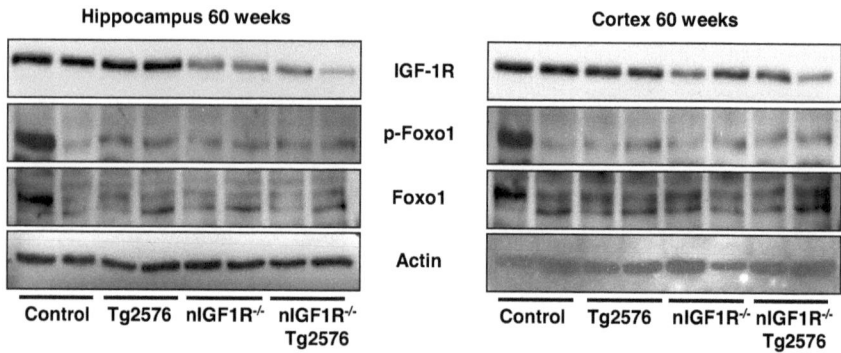

Fig. 3-32 Western blot analysis of Foxo1
Western blot analysis of IGF-1R, phospho-Foxo1 (Ser^{256}), Foxo1 and actin (loading control) protein expression in Hippocampus and cortex lysates from 60 weeks old wild type, Tg2576, nIGF-1R$^{-/-}$ and nIGF-1R$^{-/-}$Tg2576 mice. 100µg of protein were applied on 10% SDS-PAGE. Examples of 2 independent experiments.

3.8.2 Investigation of the APP processing

The results of the 28 weeks old mice revealed a 50% reduction of CTFs and Aβ$_{1-40}$ peptides. Consequently WB analyses of enzymes involved in APP processing of 60 weeks old mice were performed as well. As presented in Figure 3-33 and similarly to the results of the analysis at 28 weeks no changes were detected in expression of amyloid precursor protein (Holo-APP) of the appropriate genotype. The occurrence of CTFs in the hippocampus of nIGF-1R$^{-/-}$Tg2576 animals compared to Tg2576 mice were decreased as well. No differences were detected in the cortex of Tg2576 and nIGF-1R$^{-/-}$Tg2576 mice.

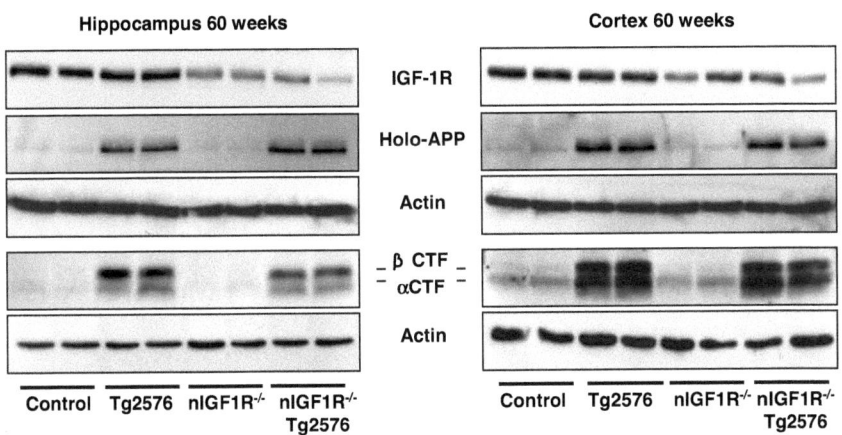

Fig. 3-33 Western blot analysis of C-teminal fragments (CTFs)
Western blot analysis of IGF-1R, Holo-APP, α- and β-C-terminal fragments protein expression in Hippocampus and cortex lysates from 60 weeks wild type, Tg2576, nIGF-1R$^{-/-}$ and nIGF-1R$^{-/-}$Tg2576 mice. 100μg of protein were applied on 10% SDS-PAGE or 15% SDS-PAGE (CTFs). Examples of 2 independent experiments.

To quantify Aβ$_{1-40/42}$ peptides concentration ELISAs were performed. Similar to the results of 28 weeks old animals a downregulation of Aβ$_{1-40}$ peptides in nIGF-1R$^{-/-}$Tg2576 animals to approximately 50% was detected compared to Tg2576, while no differences was detected in the cortex. Moreover 50% decreases of Aβ$_{1-42}$ peptide in hippocampi of nIGF-1R$^{-/-}$Tg2576 mice were detected in comparison with Tg2576 mice. In the cortex no changes were detected.

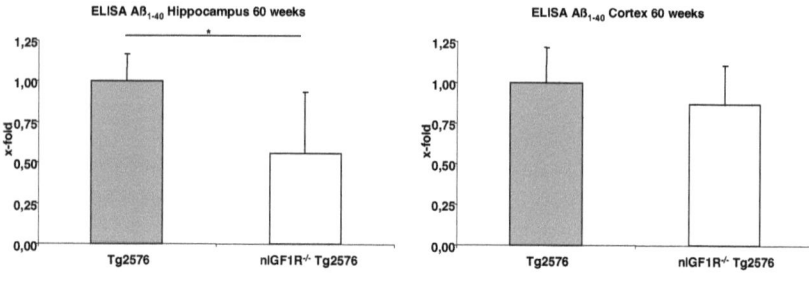

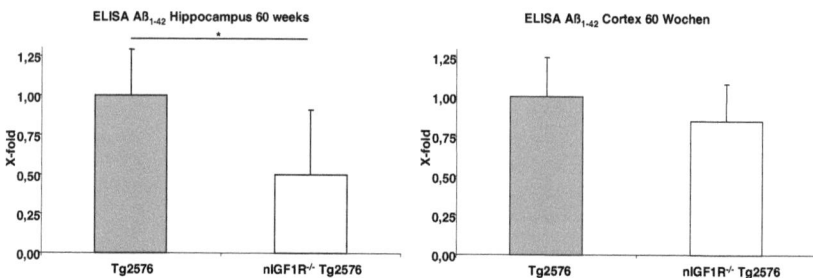

Fig. 3-34 Quantification of Aβ$_{1-40/42}$ in 60 weeks old Tg2576 and nIGF-1R$^{-/-}$Tg2576
ELISA analysis of Aβ$_{1-40}$ and Aβ$_{1-42}$ in hippocampus and cortex from 60 weeks old Tg2576 (grey bars) and nIGF-1R$^{-/-}$Tg2576 (white bars) mice. Values are means ± SD, n = 6, * unpaired Student's t-test p-value ≤ 0,05.

To visualize the amyloid plaque deposition thioflavin-S staining was performed. Thioflavin binds beta sheets, such as those in amyloid oligomers and undergoes a shift of its excitation spectrum resulting in a fluorescence signal. As seen in Figure 3-35 the plaque burden in nIGF-1R$^{-/-}$Tg2576 (upper panels) animals was lower compared to Tg2576 animals (lower panels). In addition it seems to be that plaque size in Tg2576 was larger than in nIGF-1R$^{-/-}$Tg2576.

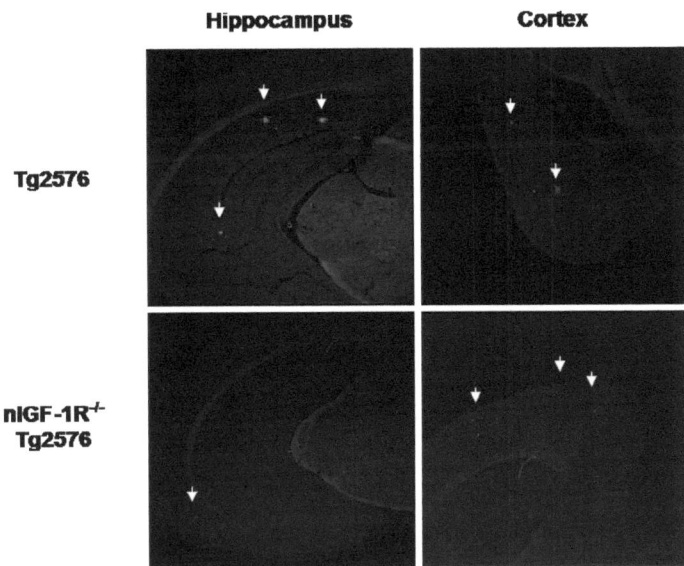

Fig. 3-35 Histochemical staining of Amyloid plaques I
Thioflavin-S immunohistochemical staining of dissected brains of 60 weeks old Tg2576 and nIGF-1R$^{-/-}$Tg2576. White arrows indicate Aβ plaque deposition composed of beta sheets. 40x magnification; green: FITC= Thioflavin-S stained beta sheets

Higher magnification of thioflavin-S stained brain sections reveals different plaque morphology in Tg2576 and nIGF-1R$^{-/-}$Tg2576 animals. Plaques of Tg2576 animals appeared to be larger and more diffuser whereas plaques of nIGF-1R$^{-/-}$Tg2576 are smaller and denser (cp. Figure 3-35).

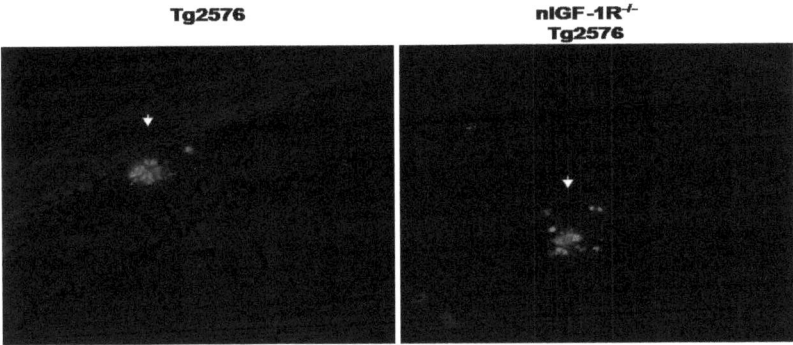

Fig. 3-36 Histochemical stainings of Amyloid plaques II
Monochrome picture of thioflavin-S immunohistochemical staining of cortex of 60 weeks old Tg2576 and nIGF-1R$^{-/-}$Tg2576. White arrows indicate Aβ plaque deposition composed of beta sheets. 100x magnification; green: FITC= Thioflavin stained beta sheets.

The previous results of APP processing revealed reduced CTFs and Aβ protein level in himppocampi of nIGF-1R$^{-/-}$Tg2576 animals compared to Tg2576. Furthermore plaque burden were reduced in nIGF-1R$^{-/-}$Tg2576 animals. Thus nIGF-1R$^{-/-}$ might have an influence on Aβ production or clearance. Consequently proteins responsible for clearance like IDE, ApoE, Neprylisin and α2 macroglobulin (α2M) were analysed. Figure 3-37 reveals no changes in the protein expression of these proteins. Consequently the proteins involved in the production of Aβ were examined.

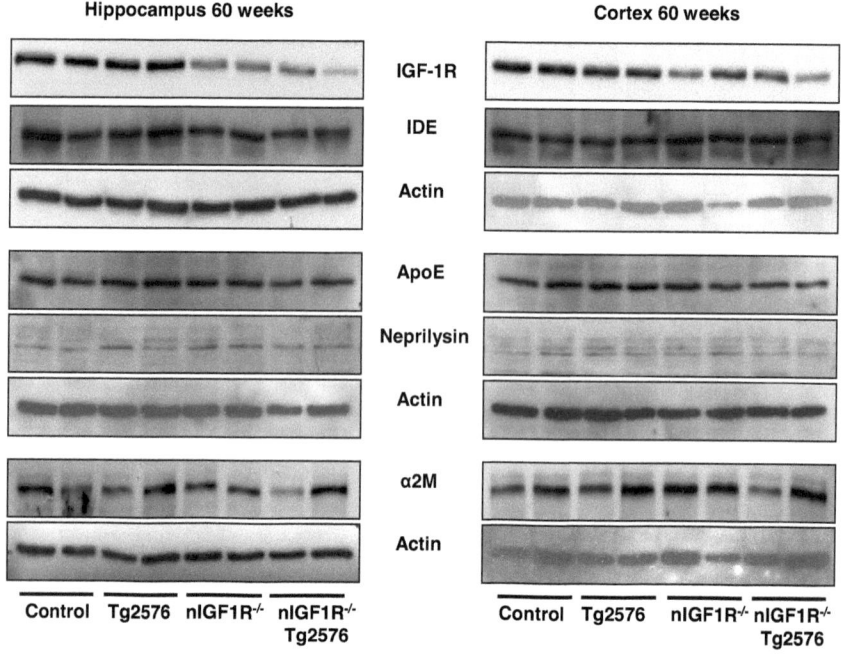

Fig. 3-37 Western blot analysis of proteins involved in clearance of Aβ
Western blot analysis of IGF-1R, IDE, ApoE, Neprilysin, α2macroglubolin (α2M) and actin (loading control) protein expression in Hippocampus and cortex lysates from 60 weeks old wild type, Tg2576, nIGF-1R$^{-/-}$ and nIGF-1R$^{-/-}$Tg2576 mice. 100μg of protein were applied on 10% SDS-PAGE. Examples of 2 independent experiments.

Cleavage of APP by α- or β-secretases directly leads to production of α- or β-CTFs. As indicated in Figure 3-37 no changes were observed in the expression of the putative α-seretase ADAM-10 or ADAM-17 (TACE) respectively. Similarly no changes were seen for the β-secretase BACE-1. No changes in protein expression of presenilin -1 were detected as well.

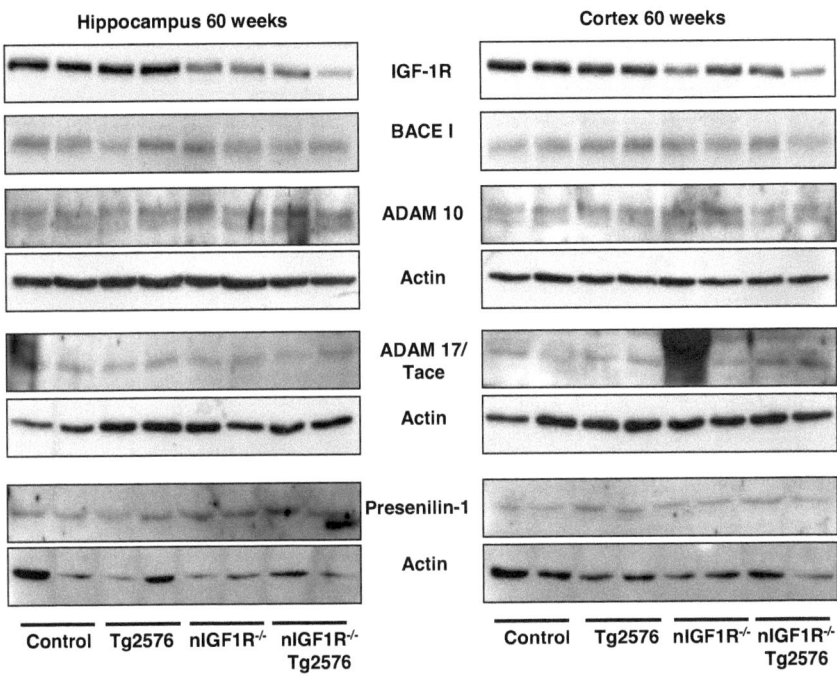

Fig. 3-38 Western blot analysis of α-, β- and γ-secretases in hippocampus and cortex of 60 weeks old mice
Western blot analysis of IGF-1R, BACE-1, ADAM-10 (active form 60kDa), ADAM-17 (TACE) (active form 85kDa), presenilin-1 and actin (loading control) protein expression in Hippocampus and cortex lysates from 60 weeks old wild type, Tg2576, nIGF-1R$^{-/-}$ and nIGF-1R$^{-/-}$Tg2576 mice. 100µg of protein were applied on 10% SDS-PAGE. Examples of 2 independent experiments.

As demonstrated for the proteins responsible for clearance no changes were seen in the protein expression of the different secretases, which are involved in the production of CTFs and Aβ by processing APP. Thus, neither the amount of proteins involved in clearance nor the amount of Aβ producing enzymes could offer an explanation for the diminished CTFs and Aβ peptides. For that reason α- and β-secretase activity assays were peformed. Due to the overexpression of APP^{sw} in

Tg2576 abundant APP^{sw} compete with the labelled substrates of the performed secretase activity assays, therefore no reliable results from lysates of Tg2576 and nIGF-1R$^{-/-}$Tg2576 were expected. Therefore, WT and nIGF-1R$^{-/-}$ mice were further analysed.

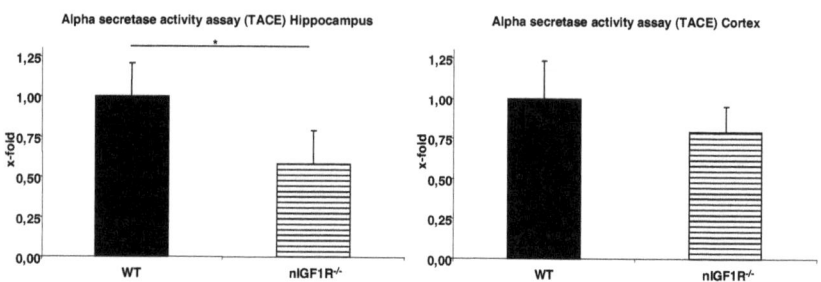

Fig. 3-39 α-secretase activity assay
α-secretase activity measurement in hippocampus and cortex lysates of 60 weeks old wild type (black bars) and nIGF-1R$^{-/-}$ (black and white striped bars) Data represented mean ± SD (n ≥ 4). unpaired Student's t-test p-value ≤ 0,05.

The activity of α-secretase in hippocampi of WT and nIGF-1R$^{-/-}$ animals expose a significantly reduced activity of 40% in nIGF-1R$^{-/-}$ mice compared to WT. In the cortex no significant changes were detected.

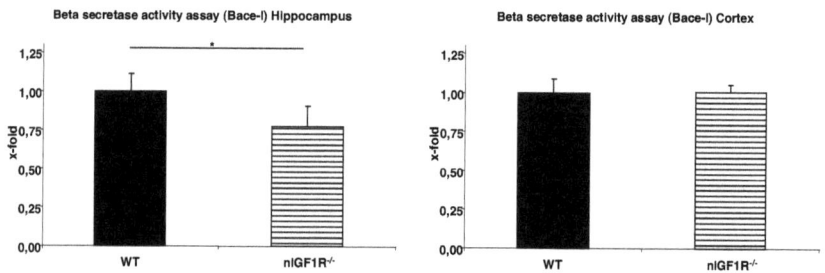

Fig. 3-40 β-secretase activity assay
β-secretase activity measurement in hippocampus and cortex lysates of 60 weeks old wild type (black bars) and nIGF-1R$^{-/-}$ (black and white striped bars) Data represented mean ± SD (n ≥ 4). unpaired Student's t-test p-value ≤ 0,05.

BACE-1 activity assays reveal similar results as the α-secretase assay but less peonounced. A 23% reduction of BACE-1 activity was detected in nIGF-1R$^{-/-}$ in the hippocampus. Measurement in the cortex did not show any differences between the genotypes.

Summarizing the results of neuron-specific IGF-1R deletion in Tg2576 background revealed several novel findings.
i) IGF-1R deletion predominantly in the hippocampus reverses APP^{sw} induced mortality;
ii) IGF-1R deletion reduces Aβ accumulation and amyloid plaque burden;
iii) IGF1R mediated signals influence APP processing due to regulation of α- and β-secretases activity.

4 Discussion

AD is a chronic progressive neurodegenerative disorder leading to loss of cognitive abilities resulting in death after an average of 8–10 years after diagnosis[175]. Growing evidence has implicated insulin and insulin-like growth factor-1 signaling as being involved in the pathogenesis of AD. Recent reports suggest that type 2 diabetes mellitus (T2DM) is a risk factor for AD. However, the underlying cellular mechanisms for this association are still unknown[176,177,178,179]. Insulin receptor and insulin-like growth factor-1 receptor signaling is markedly disturbed in the central nervous system (CNS) of AD patients[107,180,181]. Post mortem investigations of brains from patients with AD revealed a markedly down regulated expression of IR, IGF-1R, and insulin receptor substrate (IRS) proteins[102,182] and these changes progress with severity of neurodegeneration. One common feature in neurons from AD patients is a downregulation of IRS proteins and IGF-1R[102,107]. To elucidate the importance of IGF-1R signaling in the pathogenesis of AD, neuron-specific IGF-1R deleted mice (nIGF-1$^{-/-}$) were crossed with mice expressing the Swedish mutation of human APP[695] containing the double mutation Lys^{670} → Asn, Met^{671} → Leu which was found in a Swedish family with early-onset AD (APP^{SW}, Tg2576 mice). Survival, biochemical and histophathological analysis of the offsprings during an observation period of 60 weeks revealed several novel insights into the interaction of neuronal IGF-1 resistance and the pathophysiology of AD.

4.1 Tg2576 mouse model and neuron-specific IGF1-R deletion

Alzheimer's disease (AD) is pathologically characterized by senile plaques, largely composed of extracellular deposits of Aβ peptides that arise from proteolytic cleavage of APP, and neurofibrillary tangles (NFTs), composed of intracellular filamentous aggregates of hyperphosphorylated tau protein. Mouse APP is less amyloidogenic and therefore it is inpossible to analyse the development and progression of AD with its associated hallmarks, Aβ plaques and NFTs, in non transgenic mice. The Tg2576 mouse model overexpresses the 695-amino acid isoform of human amyloid precursor protein additionally harbouring a double mutation Lys^{670} →Asn, Met^{671}→ Leu. This leads to fivefold increase in Aβ$_{1-40}$ and a 14-fold increase in Aβ$_{1-42}$ accompanied by age dependent behavioral deficits[183]. However, these mice lack to develop NFTs[184]. As a result the Tg2576, a well established AD model, was used to analysing the impact of APP^{SW} and its cleavage products on development and progression of AD. In contrast, it is less applicable to investigate tau hyperphosphorylation that ends in NFTs. Therefore analysis of tau and the occurrence of NFTs were not investigated in the present study.

To investigate IGF-1 resistance in that particular AD mouse model Cre recombinase expressing mice driven by the synapsin-1 promoter (synCre) were crossed with mice carrying floxed exon-3 of the IGF-1R to generate neuron-specific IGF-1R knockout mice. To avoid a described germline deletion which has been described in male synCre mice only female synCre mice were used for breeding[185]. It was described that synpsin-1 is expressed in neurons of the CNS among others in cerebellum[169]. Therefore cerebellular granular cells from wild type and nIGF-1R$^{-/-}$ mice were generated and cultured. Unfortunately no downregulation of the IGF-1R was detected. Consequently the synCre mice were crossed with lacZ reporter mice. β-galactosidase stainings of synCre x lacZ reporter mice reveal Cre recombinase activity mainly in the hippocampal formation, and in the frontal cortex to very low extension. Western blot analyses of generated nIGF-1R$^{-/-}$ mice confirmed these data on protein level. Peripheral tissues showed no IGF-1R deletion. Therefore not only a neuron-specific but also a region specific IGF-1 resistance was further analysed.

4.2 Metabolic characterisation

Until now little is known about the role of IGF-1R/IR signaling (IIS) in the CNS. However, whole body knockout of IRS-2, a downstream target of IGF-1R, results in hyperglycemia in male mice and results in a type 2 diabetes phenotype[186]. Furthermore, up to 80% of pure C57BL/6 mice develop spontaneous hyperglycemia in the first 6 month of age possibly influencing longevity[187,188]. Therefore the influence of disturbed IIS in the CNS on glucose metabolism was investigated using GTT, ITT and long term monitoring of random fed glucose levels. No changes were detected on blood glucose levels during the observation period. Accordingly, insulin sensitivity and glucose tolerance displayed no alteration in the performed ITT and GTT. Thus an influence of IGF-1R deletion in nIGF-1R$^{-/-}$ mice on glucose metabolism is excluded. Previous studies in APP overexpressing mice have shown that lethality of these mouse models is influenced by the genetic background[166]. Since nearly all Tg2576 mice on a pure C57BL/6 background die within the first months of age it is impossible to investigate amyloid accumulation or IGF-1R signaling in this pure background[166]. The hybrid background, used in this thesis, made it possible to investigate APP^{sw} induced lethality, amyloid accumulation as well as IGF-1R signaling during aging in different brain regions without development of spontaneous hyperglycemia.

4.3 Somatic characterisation

In order to analyse the influences of somatic growth on survival of the study group characterisations of somatic development was performed. Body size of Tg2576 and nIGF-1R$^{-/-}$Tg2576 was reduced by 5-10% compared to body size of WT and nIGF-1R$^{-/-}$. Since there no changes in body size and body growth between Tg2576 and nIGF-1R$^{-/-}$Tg2576 altered somatic growth is excluded as possible cause for the survival benefit of nIGF-1R$^{-/-}$Tg2576. The reduced body size goes along with a reduced body weight in Tg2576 and nIGF-1R$^{-/-}$Tg2576. In agreement, reduced body weight of Tg2576 mice has been described in the work of Toyama et al. and has been observed the APP23 mouse model as well[189,190]. Body composition analysis of 28 and 60 weeks old mice confirm the results of body weight revealing 30-50% decreased fat content in Tg2576 mice. During aging an increasing fat proportion is normal and observed in wild type, nIGF-1R$^{-/-}$ and female nIGF-1R$^{-/-}$Tg2576 animals. Surprisingly this development was not detected in male Tg2576 and nIGF-1R$^{-/-}$Tg2576 mice. In comparison to the evaluation at 28 weeks the fat content of female Tg2576 was hardly altered and even reduced in male Tg2576 as well as in nIGF-1R$^{-/-}$Tg2576 animals. Since food intake measurements were not performed it is not excluded that Tg2576 mice eat less during aging in the present study. However, previous investigation showed no difference in food intake but an increased activity in 17 months old Tg2576 mice[191]. The APP23 mouse model displayed no reduced food intake but rather a slight increased food intake compared to wild type animals[190]. Since more than 60% of all Tg2576 animals died within 60 weeks and thus only the population with reduced phenotype survives, it could be speculated that the reduced body fat content due to an increased metabolism mimics caloric restriction leading to increased survival. Caloric restriction attenuates β-amyloid neuropathology in Tg2576 mice and is known to increase survival in different species[192,193,194,195].

Deletion of IGF-1R in neuron and glia cells has a tremendous effect. Mice lacking total brain IGF-1R develop a microcephalon, severe growth retardation, infertility, and abnormal behaviour[170].

Interestingly, brains of 60 weeks old Tg2576, nIGF-1R$^{-/-}$ and nIGF-1R$^{-/-}$Tg2576 weighed less as wild type brains. However, calculated brain-body ratios revealed only minor changes. At the end only significant differences were seen by comparison female and male wild type and Tg2576. These changes are mainly due to the relatively high differences in body weight of WT and Tg2576 animals.

As a result of the metabolic and somatic characterisation Tg2576 and nIGF-1R$^{-/-}$Tg2576 mice display no significant differences in growth, glucose metabolism and brain-body ratio. Thus, the survival benefit of nIGF-1R$^{-/-}$Tg2576 mice due to metabolic or somatic alteration is excluded.

4.4 Survival and aging

The here presented survival data of wild type, Tg2576, nIGF-1R$^{-/-}$ and nIGF-1R$^{-/-}$Tg2576 mice demonstrate that unexpectedly, neuron-specific IGF-1 resistance in the CNS in the absence of metabolic disorders prevents mortality as the most dramatic endpoint of experimental AD, in Tg2576 mice. Furthermore, heterozygosity for IGF-1R is not sufficient to rescue the APP^{sw} induced premature death in the current study. In fact, the present observation that neuron and even hippocampus IGF-1 resistance protects from AD-associated mortality argues for an evolutionary conserved life-extending mechanism from *C. elegans* to mice. In *C. elegans* neuronal DAF-2, an insulin/insulin-like growth factor receptor homolog gene, controls longevity[196]. In *Drosophila* a null mutation in CHICO, the homolog of the vertebrate IRS gene family, leads to an increase of median lifespan in heterozygous flies up to 31 % and in homozygous flies up to 48 %. The survival benefit was accompanied by decreased body size[197]. A partial loss-of-function mutation of the *Drosophila* insulin receptor (dIR) also increases lifespan up to 85 % but leads to dwarfism as well[198]. To elucidate the components of the insulin/IGF-1 receptor signaling pathway in *Drosophila* which are involved in regulation of longevity and aging various studies were performed. Overexpression of dFOXO as downstream target of IR/IGF-1R signaling increased median lifespan of female flies by up to 50 % whereas males were unaffected[199].

Whole body IGF-1R deficiency in mice results in 50 % smaller body size and these mice die after a few days due to severe developmental defects[200]. In contrast, whole body IR deficiency leads to normal birth size, but the animals die within the first hours after birth due to severe hyperglycemia and ketoacidosis[201]. Deductive, mice with deleted IR/IGF-1R signaling in whole body have a short lifespan. In contrast, heterozygous deficiency of the IGF-1R increases lifespan of female mice by up to 26 % without any effect in male animals[202]. Obviously, the organism is not capable of living completely without IRs or IGF-1Rs but partial deficiency seems to be beneficial for extending lifespan. Investigations of the downstream targets of IR/IGF-1R namely the IRS proteins displayed conflicting results. Selman et al. showed in 2008 that IRS-1 deficient mice have an 18 % increased lifespan, contrarily IRS-2 deficient mice were short-lived. Moreover, inconsistent effects were described in mice heterozygous for either IRS-1 or IRS-2[203]. Whereas one study reported an unchanged lifespan other investigators found an increased lifespan in IRS-2 heterozygous mice[203]. Heterozygoty IGF-1R and CNS restricted IRS-2-deficiency is lifespan extending in mice[170,204]. Taken together the available data suggest that the IR/IGF-1R signaling pathway is important for survival conserved in different species. Furthermore, the results of the present thesis suggesting that IIS also controls neurodegenerative disease associated lethality in rodents.

4.5 Biochemical analysis of the IGF-1R/IR signaling and APP metabolism

As described above IGF-1R deficiency in the hippocampus protects from AD associated premature death. Although no changes in protein expression involved in IIS as well as in Aβ production or clearance at the age of 28 weeks was found. IGF-1R signaling resistance was accompanied by 50% reduced Aβ-accumulation in nIGF-1R$^{-/-}$Tg2576 compared to Tg2576 mice. At the age of 60 weeks Aβ-levels were still markedly reduced. β-Amyloid consists of small peptides with N- and C-terminal heterogeneity i.e. Aβ$_{1-40}$ and Aβ$_{1-42}$, which are proteolytically released from APP via sequential cleavage by the β- and γ-secretases. Initial β-secretase cleavage generates a soluble fragment from the N-terminus of APP, while the C-terminal fragment (β-CTF) stays membrane bound. α-secretase cleavage leads to a membrane-bound C-terminal fragment (α-CTF)[205]. Interestingly, in nIGF-1R$^{-/-}$Tg2576 mice α- and β-CTF appear at significantly lower levels. Two possible reasons might explain the reduced Aβ-levels and CTFs in the hippocampus of nIGF-1R$^{-/-}$Tg2576 mice. On the one hand reduced Aβ and CTFs might be a result of decreased processing, or on the other hand a result of enhanced degradation. Transgenic expression of IDE or neprilysin, major clearance factors of Aβ, in the CNS reduces brain Aβ levels and prevents amyloid plaque formation and premature death in APP transgenic mice[206]. Concerning insulin/IGF-1 resistance it has been shown that IDE expression as an "amyloid degrading" enzyme is stimulated by the IR/IGF-1R cascade[207]. However, we could not detect any changes in IDE expression in our mouse models. Investigation of further clearance factors like neprilysin or apoE as well as α-2 macroglobulin revealed no distinguishable differences between the genotypes. Accordingly to these results it is unlikely that enhanced degradation causes the reduced Aβ-levels in nIGF-1R$^{-/-}$Tg2576 mice.

Chemokine receptor-2 (Ccr2) deficiency accelerates early disease progression by markedly impaired microglial activation. In Tg2576 mice deficient for Ccr2, accumulation of Aβ occurred earlier and these mice died significantly sooner compared to Tg2576[208]. In SHSY5Y cells as well as in primary cultured neurons chronic treatment with IGF-1 causes a switch from TrkA to p75NTR expression as seen in aging brains[209]. This switch might increase β-secretase activity indirectly by activation of neuronal sphingomyelinase which is responsible for the active liberation of the second messenger ceramide, which stabilize the β-secretase BACE-1 at least in SHSY5Y cells[210,211]. This process has been proposed to be responsible for IGF-1's effect on Aβ generation. However, BACE-1 expression was not altered at the age of 28 nor at the age of 60 weeks in these models, whereas the BACE-1 activity was significantly reduced in hippocampus of nIGF-1R$^{-/-}$at the age of 60 weeks. In contrast, it was not possible to detect an expression switch from TrkA to p75NTR by western blot analysis from brain lysates of nIGF-1R$^{-/-}$ or nIGF-1R$^{-/-}$Tg2576 compared to wild type and Tg2576 mice. Interestingly, reduced activity was also found for the α-secretase in the hippocampus of nIGF-1R$^{-/-}$ while the expression of ADAM-17 and ADAM-10 remains unchanged within the study group. The activity of the γ-secretase presenilin was not analysed. Hence, an influence of γ-secretase activity on Aβ-levels in nIGF-1R$^{-/-}$Tg2576 animals can not be excluded, although the protein expression of presenilin-1 did not show any differences. At least four independent reports linked β-amyloid accumulation to survival of APP overexpressing mice[208,212,213,214,215]. The here presented data suggest that the reduced levels of β-CTFs and, consequently of Aβ in the hippocampus of nIGF-1R$^{-/-}$Tg2576 mice due to a reduced processing is caused by a decreased β-secretase activity. Thus, the reduced amount of Aβ might be responsible for the decreased mortality in the IGF-1 resistant AD mouse model. In addition to the reduced Aβ, a diminished plaque burden as well as changes in plaque morphology were detected in nIGF-1R$^{-/-}$Tg2576 mice. It seems as if reduced IGF-1R signaling not only decelerate the processing but also modifies the Aβ plaque morphology that possibly result in a survival benefit.

Similar results have been described in *C. elegans*. Impaired insulin/IGF-1-like signaling in *C. elegans* reduces Aβ-proteotoxicity by a Foxo-dependent as well as a Foxo-independent mechanism[216]. However, no changes were observed in the expression of Foxo1 as well as in most key players of IIS like AKT and ERK-1/2. This was true not only for the unphosphorylated but also for the phosphorylated forms. Likewise, a change in protein expression of IRS-2 one of the first downstream targets of IGF-1R signaling could not be shown. Surprisingly and in contrast to IRS-2 a tremendous downregulation of IRS-1 was detected in the hippocampus of nIGF-1R$^{-/-}$ and nIGF-1R$^{-/-}$Tg2576 mice at the age of 60 weeks suggesting a distinctive role of IIS in regulating IRS protein expression during aging. Mice lacking IRS-1 show profound growth retardation and insulin resistance[217]. In contrast, mice lacking IRS-2 have mild growth defects but develop diabetes owing to a combination of insulin resistance and pancreatic β-cell dysfunction[218]. A tissue-specific role of IRS-1 and IRS-2 in IGF-1/insulin signaling has been shown in mice with mutations in IRS-1, and IRS-2[219]. Experiments performed with a β-cell line derived from IRS1$^{-/-}$ mice revealed that insulin stimulation fails to elevate cytosolic Ca^{2+} in these IRS-1–deficient cells. In contrast insulin evokes release of intracellular cell stores of Ca^{2+} in wild-type transformed β-cells[220]. Overexpression of IRS-1 increases cytosolic Ca^{2+} levels due to inhibition of uptake by the endoplasmic reticulum[221]. Therefore downregulation of IRS-1 in nIGF-1R$^{-/-}$ and nIGF-1R$^{-/-}$Tg2576 might lead to a disturbed Ca^{2+} homeostasis and might influences the activity of the secretases. Glycosylation is an important step of APP processing and changes in the cytosolic homeostasis might lead to a modified glycosylation of APP and in consequence to altered processing. To evaluate the role of ISS in APP processing further experiments are necessary. Embryonic fibroblasts and 3T3 cell lines derived from IRS-1-deficient embryos exhibit no IGF-1-stimulated IRS-1 phosphorylation or IRS-1-associated phosphatidylinositol 3-kinase (PI-3 kinase) activity but unaltered activation of the mitogen-activated protein kinases ERK-1/2[222]. Whereas the steady state of phosphorylated ERK-1/2 in

Tg2576 and nIGF-1R$^{-/-}$Tg2576 animals was not altered, markedly diminished phosphorylation of GSK-3α/β, a downstream target of PI-3 kinase, was detected in the hippocampus of nIGF-1R$^{-/-}$ and nIGF-1R$^{-/-}$Tg2576. On the other hand GSK-3 is a kinase supposed to be involved in the regulation of the secretases[223,224]. Reduced phosphorylation (Ser$^{21/9}$) of GSK-3α/β (the inaktive form of GSK-3) should result in more GSK-3 activity. The results of nIGF-1R$^{-/-}$ and nIGF-1R$^{-/-}$Tg2576 animals might suggest that increased GSK-3 activity regulates the processing of APP by slowing down activity of the secretases. In contrast to our observation results an inhibition of GSK-3α/β by lithium to a decreased Aβ production in CHO$_{APPsw}$ cells[224]. It has to be noted that the downregulation of phospho-GSK-3β (Ser9) was not observed at an age of 28 weeks despite reduction of CTFs and Aβ. To elucidate the mechanism of Aβ accumulation further studies need to be performed.

The present survival studies and biochemical investigation reveal the importance of IGF-1R signaling for premature death caused by APP^{sw} overexpression. Downregulation of IGF-1R in the hippocampus counter the premature death in Tg2576 mice. In addition Aβ- and CTF-levels are reduced as consequences of a decelerated α- and β-secretase activity in response to IGF-1R signaling resistance.
Taken together the present thesis reveal several novel findings i) neuronal IGF-1R deficiency protects against APP^{sw}-induced lethality, ii) deletion of IGF-1R mediated signals reduces Aβ accumulation in mice via decelerated β-secretase activity .
Thus, downregulation of IGF-1R observed in neurons of patients suffering from Alzheimer's disease is most likely a compensatory phenomenon to decrease amyloid burden and prolong survival.

4.6 Perspectives and experimental approach

The rescue of premature death in Tg2576 mice observed in nIGF-1R$^{-/-}$Tg2576 animals is most likely a result of the decreased Aβ-accumulation caused by the decelerated secretase activity. At the moment little is known about the exact mechanism. In current work only a correlation between modified IGF-1R signaling cascade and reduced activity of the secretases is discribed. To get more information about the underlying mechanism neuroblastoma cells with chronically alter IGF-1R signaling can be used. Since in SHSY5Y cells IGF-1R signaling is mainly mediated via IRS-2, the cells could be modify in a way that one cellline will be overexpressing

IRS-2, one will express a siRNA knocking down IRS-2 and one will be stably transfected with mutated siRNA as control. In these cell lines α-, β- and γ-secretase activity can further be investigated. Furthermore, the role of IGF-1R mediated signals for APP trafficking can be analysed using the above mentioned cell lines, transfected with a tagged APP as well. In addition different kinase inhibitors of the IGF-1R downstream signaling cascade might be used to address the specific signaling pathways involved in APP processing and secretase regulation.

These experiments will reveal the molecular mechanism underlying the effect of IGF-1R signaling on APP processing and clearance.

5 Summary

Post mortem investigations of brains from patients with AD revealed a markedly down regulated expression of IGF-1R, and insulin receptor substrate (IRS) proteins, and these changes progress with severity of neurodegeneration. To investigate the role of neuronal IGF-1R signaling in AD neuron-specific IGF-1R (nIGF-1R$^{-/-}$) deficient mice were generated. These mice were crossed with mice expressing the Swedish mutation of human APP695 harbouring the double mutation Lys^{670}→ Asn, Met^{671}→ Leu which was found in a Swedish family with early-onset AD (APP^{sw}, Tg2576 mice). nIGF-1R$^{-/-}$ mice were generated using the cre-loxP-system under the control of the neuron-specific synapsin-1 promoter and crossed them into the Tg2576 background. The offsprings of these mice (WT, Tg2576, nIGF-1R$^{-/-}$, nIGF-1R$^{-/-}$Tg2576) were analysed at two different time points. Kaplan-Meier analysis, amyloid accumulation as well as metabolic and somatic factors of the offspring were investigated during an observation period of 60 weeks. Western blot analysis of isolated hippocampi displayed a 40% reduced IGF-1R expression in nIGF-1R$^{-/-}$ and nIGF-1R$^{-/-}$Tg2576 compared to WT and Tg2576 animals, whereas other brain regions e.g. cortex or cerebellum did not show significant IGF-1R deletion. Thus, conditional IGF-1R deletion using Cre recombinase expression under the control of the synapsin-1 promoter leads to a hippocampus-specific downregulation of IGF-1R. Further analysis of Cre recombinase expression in a lacZ reporter mouse strain revealed a Cre recombinase activity driven by the synapsin-1 promoter in the dentus gyrus and the CA3 region of the hippocampus. Kaplan-Meier-analysis revealed a 60% mortality of Tg2576 mice after 60 weeks of observation. In contrast nIGF-1R$^{-/-}$Tg2576 were protected against premature mortality of Tg2576 mice (p≤0.02; Tg2576 vs. nIGF-1R$^{-/-}$ Tg2576). Isolated hippocampi of 28 and 60 weeks old nIGF-1R$^{-/-}$Tg2576 animals showed a 50% reduced Aβ_{1-40} and Aβ_{1-42} accumulation compared to Tg2576. Additionally, APP α- and β-C-terminal fragments were reduced in hippocampi of nIGF-1R$^{-/-}$Tg2576 compared Tg2576 mice due to a modification of α- and β-secretases activity. In addition Aβ plaque burden was reduced in Tg2576 animals with neuronal IGF-1R deletion.

Taken together the results of the present thesis demonstrate that decreased neuronal IGF-1R signaling predominantly in the hippocampus protects against APP^{sw} induced mortality. Moreover IGF-1R mediated signals influence APP processing due to a modification of α- and β-secretases leading to reduced Aβ accumulation and amyloid plaque burden. Thus, downregulation of IGF-1R observed in neurons of patients suffering from Alzheimer's disease is most likely a compensatory phenomenon to decrease amyloid accumulation and prolong survival.

6 References

[1] Maurer Ulrike, Maurer Konrad (2003). Alzheimer: the life of a physician and the career of a disease. New York: Columbia University Press. pp. 270. ISBN 0-231-11896-1.
[2] Walsh JS, Welch HG, Larson EB. Survival of outpatients with Alzheimer-type dementia. Ann Intern Med 1990; 113:429–34.
[3] Burns A, Jacoby R, Levy R. Psychiatric phenomena in Alzheimer's disease. Br J Psychiatry 1990;157:72-94.
[4] American Psychiatric Association (2000). Diagnostic and Statistical Manual of Mental Disorders, Fourth edition. Text Revision (DSM-IV-TR®).
[5] McKhann, G., Drachman, D., Folstein, M., Katzman, R., Price, D., & Stadlan, E. M. (1984). Clinical diagnosis of Alzheimer's disease: report of the NINCDS-ADRDA Work Group under the auspices of Department of Health and Human Services Task Force on Alzheimer's disease. Neurology 34, 939–944.
[6] Dubois B, Feldman HH, Jacova C, et al (August 2007). "Research criteria for the diagnosis of Alzheimer's disease: revising the NINCDS-ADRDA criteria". Lancet Neurol 6 (8): 734–46.
[7] Rocca WA, Hofman A, Brayne C, Breteler MM, et al. Frequency and distribution of Alzheimer's disease in Europe: a collaborative study of 1980 –1990 prevalence findings. The EURODEM-Prevalence Research Group. Ann Neurol 1991;30:381–390.
[8] Campion D, Dumanchin C, Hannequin D, Dubois B, et al. Early-onset autosomal dominant Alzheimer disease: prevalence, genetic heterogeneity, and mutation spectrum. Am J Hum Genet 1999;65:664–670.
[9] Lindsay, J. et al. Risk factors for Alzheimer's disease: a prospecave analysis from the Canadian Study of Health and Aging. Am J Epidemiol 156, 445-453 (2002).
[10] Bachman, D. L. et al. Incidence of demenaa and probable Alzheimer's disease in ageneral populaaon: the Framingham Study. Neurology 43, 515-519 (1993).
[11] Farrer LA, O'Sullivan DM, Cupples LA, Growdon JH, et al. Assessment of genetic risk for Alzheimer's disease among first-degree relatives. Ann Neurol 1989;25:485–493.
[12] Silverman JM, Li G, Zaccario ML, Smith CJ, et al. Patterns of risk in first-degree relatives of patients with Alzheimer's disease. Arch Gen Psychiatry 1994;51:577–586.
[13] Farrer LA, O'Sullivan DM, Cupples LA, Growdon JH, et al. Assessment of genetic risk for Alzheimer's disease among first-degree relatives. Ann Neurol 1989;25:485–493.
[14] Tschanz, J. T. et al. Demenaa: the leading predictor of death in a defined elderly population: the Cache County Study. Neurology 62, 1156 -1162 (2004).
[15] Bermejo-Pareja F, Benito-León J, Vega S, Medrano MJ, Román GC (January 2008). "Incidence and subtypes of dementia in three elderly populations of central Spain". J. Neurol. Sci. 264 (1–2): 63–72.
[16] Di Carlo A, Baldereschi M, Amaducci L, et al (January 2002). "Incidence of dementia, Alzheimer's disease, and vascular dementia in Italy. The ILSA Study". J Am Geriatr Soc 50 (1): 41–8.
[17] Di Carlo A, Baldereschi M, Amaducci L, et al (January 2002). "Incidence of dementia, Alzheimer's disease, and vascular dementia in Italy. The ILSA Study". J Am Geriatr Soc 50 (1): 41–8.
[18] Hebert LE, Scherr PA, Bienias JL, Bennett DA, Evans DA. Alzheimer disease in the US population: prevalence estimates using the 2000 census. Arch Neurol 2003;60: 1119–1122.
[19] Ernst RL, Hay JW. Economic research on Alzheimer disease: a review of the literature. Alzheimer Dis Assoc Disord 1997;11(suppl 6):135–145.
[20] Tanzi RE, Gusella JF, Watkins PC, et al. Amyloid beta-protein gene—cDNA, messenger-RNA distribution, and genetic-linkage near the Alzheimer locus. Science 1987; 235:880–4.
[21] Goate AM. Molecular genetics of Alzheimer's disease. Geriatrics 1997;52:S9–17.
[22] Campion D, Brice A, Dumanchin C, et al. A novel presenilin-1 mutation resulting in familial Alzheimer'sdisease with an onset age of 29 years. Neuroreport 1996;7:1582–4.
[23] Van Broeckhoven C. Presenilins and Alzheimer-disease. Nat Genet 1995;11:230–2.
[24] Levitan D, Greenwald I. Facilitation of lin-12-mediated signaling by Sel-12, a Caenorhabditis-elegans S182 Alzheimer's-disease gene. Nature 1995;377:351–4.
[25] Wong PC, Zheng H, Chen H, et al. Presenilin 1 is required for Notch1 DII1 expression in the paraxial mesoderm. Nature 1997;387:288–92.

[26] Shen J, Bronson RT, Chen DF, et al. Skeletal and CNS defects in presenilin-1-deficient mice. Cell 1997;89:629– 39.
[27] Levy-Lahad E,Wascow, Poorkaj P, et al.Candidate gene for the chromosome-1 familial Alzheimer's-disease locus. Science 1995;269:973–7.
[28] Jarrett JT, Berger EP, Lansbury PT. The carboxy terminus of the beta-amyloid protein is critical for the seeding of amyloid formation—implications for the pathogenesis of Alzheimer's-disease. Biochemistry 1993;32:4693–7.
[29] Duff K, Eckman C, Zehr C, et al. Increased amyloid-beta- 42(43) in brains of mice expressing mutant presenilin-1. Nature 1996;383:710–3.
[30] Scheuner D, Eckman C, Jensen M, et al. Secreted amyloid beta-protein similar to that in the senile plaques of Alzheimer's-disease is increased in-vivo by the presenilin-1 and presenilin-2 and APP mutations linked to familial Alzheimer's-disease. Nat Med 1996;2:864–70.
[31] Citron M, Westaway D, Xia WM, et al. Mutant presenilins of Alzheimer's disease increase production of 42-residue amyloid beta-protein in both transfected cells and transgenic mice. Nat Med 1997;3:67–72.
[32] Farrer LA, Cupples LA, Haines JL, Hyman B, Kukull WA, Mayeux R, et al. Effects of age, sex, and ethnicity on the association between apolipoprotein E genotype and Alzheimer disease. A meta-analysis. APOE and Alzheimer Disease Meta Analysis Consortium. JAMA 1997;278:1349-56.
[33] Holtzman, D. M. Role of apoe/Abeta interacaons in the pathogenesis of Alzheimer's disease and cerebral amyloid angiopathy. J Mol Neurosci 17, 147-155 (2001).
[34] Cedazo-Mínguez, A. Apolipoprotein E and Alzheimer's disease: molecular mechanisms and therapeuac opportuniaes. J Cell Mol Med 11, 1227-1238 (2007).
[35] Leverenz JB, Raskind MA. Early amyloid deposition in the medial temporal lobe of young Down syndrome patients: a regional quantitative analysis. Exp Neurol 1998; 150:296 –304.
[36] Prasher VP, Farrer MJ, Kessling AM, Fisher EM, et al. Molecular mapping of Alzheimertype dementia in Down's syndrome. Ann Neurol 1998;43:380 –383.
[37] R.D. Terry, The Fine Structure of Neurofibrillary Tangles in Alzheimer's Disease, J Neuropathol Exp Neurol 22 (1963), 629–642.
[38] I. Grundke-Iqbal, K. Iqbal, M. Quinlan, Y.C. Tung, M.S. Zaidi and H.M. Wisniewski, Microtubule-associated protein tau. A component of Alzheimer paired helical filaments, J Biol Chem 261 (1986), 6084–6089.
[39] Trojanowski JQ, Lee VM, The role of tau in Alzheimer´s disease. Med Clin North Am 2002;86: 615-27.
[40] Feijoo C, Campell DG, Evidence that phosphorylation of the microtubule-associated protein tau by SAPK4/p38delta at Thr50 promotes microtubule assembly. J Cell Sci 2005; 118: 397-408.
[41] Weingarten MD, Lockwood AH, Hwo SY, Kirschner MW. A protein factor essential for microtubule assembly. Proc Natl Acad Sci USA. 1975; 72: 1858–62.
[42] Morishima-Kawashima M, Hasegawa M, Takio K, Suzuki M, Yoshida H, Titani K, Ihara Y. Prolinedirected and non-proline-directed phosphorylation of PHF-tau. J Biol Chem. 1995; 270:823–9.
[43] Hanger DP, Betts JC, Loviny TL, Blackstock WP, Anderton BH. New phosphorylation sites identified in hyperphosphorylated tau (paired helical filamenttau) from Alzheimer's disease brain using nanoelectrospray mass spectrometry. J Neurochem. 1998; 71: 2465–76.
[44] Singh TJ, Grundke-Iqbal I, McDonald B, Iqbal K. Comparison of the phosphorylation of microtubuleassociated protein tau by non-proline dependent protein kinases. Mol Cell Biochem. 1994; 131: 181–9.
[45] Gong CX, Lidsky T, Wegiel J, Zuck L, Grundke-Iqbal I, Iqbal K. Phosphorylation of microtubuleassociated protein tau is regulated by protein phosphatase 2A in mammalian brain. Implications for neurofibrillary degeneration in Alzheimer's disease. J Biol Chem. 2000; 275: 5535–44.
[46] Liu F, Grundke-Iqbal I, Iqbal K, Gong CX. Contributions of protein phosphatases PP1, PP2A, PP2B and PP5 to the regulation of tau phosphorylation. Eur J Neurosci. 2005; 22: 1942–50.

[47] Tanaka T, Zhong J, Iqbal K, Trenkner E, Grundke-Iqbal I. The regulation of phosphorylation of tau in SY5Y neuroblastoma cells: the role of protein phosphatases. FEBS. Lett. 1998; 426: 248–54.
[48] Khatoon S, Grundke-Iqbal I, Iqbal K. Brain levels of microtubule-associated protein tau are elevated in Alzheimer's disease: a radioimmuno-slot-blot assay for nanograms of the protein. J Neurochem. 1992; 59: 750–3.
[49] Khatoon S, Grundke-Iqbal I, Iqbal K. Levels of normal and abnormally phosphorylated tau in different cellular and regional compartments of Alzheimer disease and control brains. FEBS Lett. 1994; 351: 80–4.
[50] Iqbal K, Grundke-Iqbal I, Smith AJ, George L, Tung YC, Zaidi T. Identification and localization of a tau peptide to paired helical filaments of Alzheimer disease.
[51] Lee VM, Balin BJ, Otvos L Jr, Trojanowski JQ. A68: a major subunit of paired helical filaments and derivatized forms of normal Tau. Science. 1991; 251: 675–8.
[52] Alonso AD, Zaidi T, Grundke-Iqbal I, Iqbal K. Role of abnormally phosphorylated tau in the breakdown of microtubules in Alzheimer disease. Proc Natl Acad Sci USA. 1994; 91:5562–6.
[53] Zheng H, Koo EH (2006) The amyloid precursor protein: beyond amyloid. Mol Neurodegener 1:5.
[54] Hung AY, Koo EH, Haass C, Selkoe DJ (1992) Increased expression of beta-amyloid precursor protein during neuronal differentiation is not accompanied by secretory cleavage. Proc Natl Acad Sci U S A 89: 9439–9443.
[55] Selkoe, D. J. (2001) Physiol. Rev. 81, 741–766.
[56] Tomita, S., Kirino, Y., and Suzuki, T. (1998) J. Biol. Chem. 273, 6277–6284.
[57] Small, S. A., and Gandy, S. (2006) Neuron 52, 15–31.
[58] Kojro, E. & Fahrenholz, F. The non-amyloidogenic pathway: structure and function of alpha-secretases. Subcell Biochem 38, 105-127 (2005).
[59] Yu C, Kim SH, Ikeuchi T, Xu H, Gasparini L, Wang R, Sisodia SS (2001) Characterization of a presenilin-mediated amyloid precursor protein carboxyl-terminal fragment g. Evidence for distinct mechanisms involved in g-secretase processing of the APP and Notch1 transmembrane domains. J Biol Chem 276: 43756–43760.
[60] Gervais FG, Xu D, Robertson GS, Villaincourt JP, Zhu Y, Huang Y, LeBlanc A, Smith D, Rigby M, Shearman MS, Clarke EE, Zheng H, Van Der Ploeg RH, Rufollo SC, Thornberry NA, Xanthoudakis S, Zamboni RJ, Roy S, Nicholson DW (1999) Involvement of caspases in proteolytic cleavage of Alzheimer's amyloid-beta precursor protein and amyloidogenic A beta peptide formation. Cell 97: 395–406.
[61] Vassar, R. et al. Beta-secretase cleavage of Alzheimer's amyloid precursor protein by the transmembrane asparac protease BACE. Science 286, 735-741 (1999).
[62] Citron, M. et al. Mutaaon of the beta-amyloid precursor protein in familial Alzheimer's disease increases beta-protein production. Nature 360, 672-674 (1992).
[63] Kimberly WT, Wolfe MS (2003) Identity and function of gamma-secretase. J Neurosci Res 74: 353–360.
[64] Sastre M, Steiner H, Fuchs K, Capell A, Multhaup G, Condron MM, Teplow DB, Haass C (2001) Presenilin-dependent gamma-secretase processing of beta-amyloid precursor protein at a site corresponding to the S3 cleavage of Notch. EMBO Rep 2: 835–841.
[65] Jarrew, J. T., Berger, E. P. & Lansbury, P. T. The carboxy terminus of the beta amyloid protein is criacal for the seeding of amyloid formation: implications for the pathogenesis of Alzheimer's disease. Biochemistry 32, 4693-4697 (1993).
[66] Daigle I, Li C (1993) apl-1, a Caenorhabditis elegans gene encoding a protein related to the human beta-amyloid protein precursor. Proc Natl Acad Sci USA 90:12045–12049.
[67] Rosen DR, Martin-Morris L, Luo LQ, White K (1989) A Drosophila gene encoding a protein resembling the human beta-amyloid protein precursor. Proc Natl Acad Sci USA 86:2478 – 2482.
[68] Wasco W, Bupp K, Magendantz M, Gusella JF, Tanzi RE, Solomon F (1992) Identification of a mouse brain cDNA that encodes a protein related to the Alzheimer disease-associated amyloid beta protein precursor. Proc Natl Acad Sci USA 89:10758 –10762.

[69] Wasco W, Gurubhagavatula S, Paradis MD, Romano DM, Sisodia SS, Hyman BT, Neve RL, Tanzi RE (1993) Isolation and characterization of APLP2 encoding a homologue of the Alzheimer's associated amyloid beta protein precursor. Nat Genet 5:95–100.
[70] Slunt HH, Thinakaran G, von Koch C, Lo AC, Tanzi RE, Sisodia SS (1994) Expression of a ubiquitous, cross-reactive homologue of the mouse beta-amyloid precursor protein (APP). J Biol Chem 269:2637–2644.
[71] De Strooper B, Simons M, Multhaup G, Van Leuven F, Beyreuther K, Dotti CG (1995) Production of intracellular amyloid-containing fragments in hippocampal neurons expressing human amyloid precursor protein and protection against amyloidogenesis by subtle amino acid substitutions in the rodent sequence. EMBO J 14:4932– 4938.
[72] Duckworth, W.C., Bennett, R.G., and Hamel, F.G. (1998). Insulin degradation: progress and potential .Endocr. Rev. *19*, 608–624.
[73] Selkoe, D.J. (2001). Clearing the brain's amyloid cobwebs. Neuron *32*, 177–180.
[74] Farris, W., Mansourian, S., Chang, Y., Lindsley, L., Eckman, E.A., Frosch, M.P., Eckman, C.B., Tanzi, R.E., Selkoe, D.J., and Guenette, S. (2003). Insulin-degrading enzyme regulates the levels of insulin, amyloid beta-protein, and the beta-amyloid precursor protein intracellular domain in vivo. Proc. Natl. Acad. Sci. USA *100*, 4162–4167.
[75] Leissring, M.A., Farris, W., Chang, A.Y., Walsh, D.M., Wu, X., Sun, X., Frosch, M.P., and Selkoe, D.J. (2003). Enhanced proteolysis of beta-amyloid in APP transgenic mice prevents plaque formation, secondary pathology, and premature death. Neuron *40*, 1087–1093.
[76] Bertram, L., and Tanzi, R.E. (2004). Alzheimer's disease: one disorder, too many genes?Hum. Mol. Genet. *13*, R135–141.
[77] Turner, A.J., Isaac, R.E., and Coates, D. (2001). The neprilysin (NEP) family of zinc metalloendopeptidases: genomics and function. Bioessays *23*, 261–269.
[78] Marr, R.A., Rockenstein, E., Mukherjee, A., Kindy, M.S., Hersh, L.B., Gage, F.H., Verma, I.M., and Masliah, E. (2003). Neprilysin gene transfer reduces human amyloid pathology in transgenic mice. J. Neurosci. *23*, 1992–1996.
[79] Zlokovic, B.V. (2004). Clearing amyloid through the blood-brain barrier. J. Neurochem. *89*, 807–811.
[80] Deane, R., Wu, Z., Sagare, A., Davis, J., Du Yan, S., Hamm, K., Xu, F., Parisi, M., LaRue, B., Hu, H.W., et al. (2004). LRP/amyloid beta-peptide interaction mediates differential brain efflux of Abeta isoforms. Neuron *43*, 333–344.
[81] Herz, J. (2003). LRP: a bright beacon at the blood-brain barrier. J. Clin. Invest. *112*, 1483–1485.
[82] Kuentzel, S. L., Ali, S. M., Altman, R. A., Greenberg, B. D. & Raub, T. J. The Alzheimer beta-amyloid protein precursor/protease nexin-II is cleaved by secretase in a trans-Golgi secretory compartment in human neuroglioma cells. Biochem J 295, 367-378 (1993).
[83] Lammich, S. et al. Consatuave and regulated alpha-secretase cleavage of Alzheimer's amyloid precursor protein by a disintegrin metalloprotease. Proc Natl Acad Sci USA 96, 3922-3927 (1999).
[84] Hussain, I., Powell, D., Howlett, D. R., Tew, D. G., Meek, T. D., Chapman, C., Gloger, I. S.,Murphy, K. E., Southan, C. D., Ryan, D. M., Smith, T. S., Simmons, D. L., Walsh, F. S., Dingwall, C., and Christie, G. (1999). Identification of a novel aspartic protease (Asp 2) as beta-secretase. Mol. Cell. Neurosci. 14, 419–427.
[85] Sinha, S., Anderson, J. P., Barbour, R., Basi, G. S., Caccavello, R., Davis, D., Doan, M., Dovey, H. F., Frigon, N., Hong, J., Jacobson-Croak, K., Jewett, N., Keim, P., Knops, J., Lieberburg, I., Power, M., Tan, H., Tatsuno, G., Tung, J., Schenk, D., Seubert, P., Suomensaari, S. M., Wang, S., Walker, D., Zhao, J., McConlogue, L., and John, V. (1999). Purification and cloning of amyloid precursor protein beta-secretase from human brain. Nature 402, 537–540.
[86] Yan, R., Bienkowski, M. J., Shuck, M. E., Miao, H., Tory, M. C., Pauley, A. M., Brashler, J. R., Stratman, N. C., Mathews, W. R., Buhl, A. E., Carter, D. B., Tomasselli, A. G., Parodi, L. A., Heinrikson, R. L., and Gurney, M. E. (1999). Membrane-anchored aspartyl protease with Alzheimer's disease beta-secretase activity. Nature 402, 533–537.

[87] Zhao, J., Paganini, L., Mucke, L., Gordon, M., Refolo, L., Carman, M., Sinha, S.,Oltersdorf, T., Lieberburg, I., and McConlogue, L. (1996). Beta-secretase processing of the beta-amyloid precursor protein in transgenic mice is efficient in neurons but inefficient in astrocytes. *J. Biol. Chem.* 271, 31407–31411.

[88] Bennett, B. D., Denis, P., Haniu, M., Teplow, D. B., Kahn, S., Louis, J.-C., Citron, M., and Vassar, R. (2000). A furin-like convertase mediates propeptide cleavage of BACE, the Alzheimer's beta -secretase. *J. Biol. Chem.* 275, 37712–37717.

[89] Iwatsubo, T. The gamma-secretase complex: machinery for intramembrane proteolysis. *Curr Opin Neurobiol* 14, 379-383 (2004).

[90] Thinakaran, G., Borchelt, D. R., Lee, M. K., Slunt, H. H., Spitzer, L., Kim, G., Ratovitsky, T.,Davenport, F., Nordstedt, C., Seeger, M., Hardy, J., Levey, A. I., Gandy, S. E., Jenkins, N. A., Copeland, N. G., Price, D. L., and Sisodia, S. S. (1996). Endoproteolysis of presenilin 1 and accumulation of processed derivatives in vivo. *Neuron* 17, 181–190.

[91] Francis, R., McGrath, G., Zhang, J., Ruddy, D. A., Sym, M., Apfeld, J., Nicoll, M., Maxwell, M., Hai, B., Ellis, M. C., Parks, A. L., Xu, W., Li, J., Gurney, M., Myers, R. L., Himes, C. S., Hiebsch, R. D., Ruble, C., Nye, J. S., and Curtis, D. (2002). aph-1 and pen-2 are required for Notch pathway signaling, gamma-secretase cleavage of betaAPP, and presenilin protein accumulation. *Dev. Cell* 3, 85–97.

[92] Goutte, C., Tsunozaki, M., Hale, V. A., and Priess, J. R. (2002). APH-1 is a multipass membrane protein essential for the Notch signaling pathway in Caenorhabditis elegans embryos. *Proc. Natl.Acad. Sci. U. S. A.* 99, 775–779.

[93] Yu, G., Nishimura, M., Arawaka, S., Levitan, D., Zhang, L., Tandon, A., Song, Y. Q., Rogaeva, E., Chen, F., Kawarai, T., Supala, A., Levesque, L., Yu, H., Yang, D. S., Holmes, E., Milman, P., Liang, Y., Zhang, D. M., Xu, D. H., Sato, C., Rogaev, E., Smith, M., Janus, C., Zhang, Y., Aebersold, R., Farrer, L. S., Sorbi, S., Bruni, A., Fraser, P., and St George-Hyslop, P. (2000). Nicastrin modulates presenilin-mediated notch/glp-1 signal transduction and betaAPP processing. *Nature* 407, 48–54.

[94] Shah, S., Lee, S. F., Tabuchi, K., Hao, Y. H., Yu, C., LaPlant, Q., Ball, H., Dann, C. E., III, Sudhof, T., and Yu, G. (2005). Nicastrin functions as a gamma-secretase-substrate receptor. *Cell* 122, 435–447.

[95] Prokop, S., Shirotani, K., Edbauer, D., Haass, C., and Steiner, H. (2004). Requirement of PEN-2 for stabilization of the presenilin N-/C-terminal fragment heterodimer within the gamma-secretase complex. *J. Biol. Chem.* 279, 23255–23261.

[96] Hasegawa, H., Sanjo, N., Chen, F., Gu, Y. J., Shier, C., Petit, A., Kawarai, T., Katayama, T., Schmidt, S. D., Mathews, P. M., Schmitt-Ulms, G., Fraser, P. E., and St George-Hyslop, P. (2004). Both the sequence and length of the C terminus of PEN-2 are critical for intermolecular interactions and function of presenilin complexes. *J. Biol. Chem.* 279, 46455–46463.

[97] Ott, A., Stolk, R. P., van Harskamp, F., Pols, H. A., Hofman, A., and Breteler, M. M. (1999) Diabetes mellitus and the risk of dementia: The Rotterdam Study. *Neurology* 53, 1937-1942.

[98] Luchsinger, J. A., Tang, M. X., Shea, S., and Mayeux, R. (2004) Hyperinsulinemia and risk of Alzheimer disease. *Neurology* 63, 1187-1192.

[99] Haan, M. N. (2006) Therapy Insight: type 2 diabetes mellitus and the risk of lateonset Alzheimer's disease. *Nat. Clin. Pract. Neurol.* 2, 159-166.

[100] Stewart, R., and Liolitsa, D. (1999) Type 2 diabetes mellitus, cognitive impairment and dementia. *Diabet. Med.* 16, 93-112.

[101] Lovestone, S. (1999) Diabetes and dementia: is the brain another site of end-organ damage? *Neurology* 53, 1907-1909.

[102] Steen, E., Terry, B. M., Rivera, E. J., Cannon, J. L., Neely, T. R., Tavares, R., Xu, X. J., Wands, J. R., and de la Monte, S. M. (2005) Impaired insulin and insulin-like growth factor expression and signaling mechanisms in Alzheimer's disease--is this type 3 diabetes? *J. Alzheimers Dis.* 7, 63-80.

[103] de la Monte, S. M., Tong, M., Lester-Coll, N., Plater, M., Jr., and Wands, J. R. (2006) Therapeutic rescue of neurodegeneration in experimental type 3 diabetes: relevance to Alzheimer's disease. *J. Alzheimers Dis.* 10, 89-109.

[104] Pilcher, H. (2006) Alzheimer's disease could be "type 3 diabetes". *Lancet Neurol.* 5, 388-389.
[105] Frolich, L., Blum-Degen, D., Bernstein, H. G., Engelsberger, S., Humrich, J., Laufer, S., Muschner, D., Thalheimer, A., Turk, A., Hoyer, S., Zochling, R., Boissl, K. W., Jellinger, K., and Riederer, P. (1998) Brain insulin and insulin receptors in aging and sporadic Alzheimer's disease. *J. Neural Transm.* 105, 423-438.
[106] Frolich, L., Blum-Degen, D., Riederer, P., and Hoyer, S. (1999) A disturbance in the neuronal insulin receptor signal transduction in sporadic Alzheimer's disease. *Ann. N. Y. Acad. Sci.* 893, 290-293.
[107] Moloney, A. M., Griffin, R. J., Timmons, S., O'Connor, R., Ravid, R., and O'Neill, C. (2008) Defects in IGF-1 receptor, insulin receptor and IRS-1/2 in Alzheimer's disease indicate possible resistance to IGF-1 and insulin signaling. *Neurobiol. Aging*.
[108] Rivera, E. J., Goldin, A., Fulmer, N., Tavares, R., Wands, J. R., and de la Monte, S. M. (2005) Insulin and insulin-like growth factor expression and function deteriorate with progression of Alzheimer's disease: link to brain reductions in acetylcholine. *J. Alzheimers Dis.* 8, 247-268.
[109] Rinderknecht E, Humbel RE 1978 The amino acid sequence of human insulin-like growth factor I and its structural homology with proinsulin. J Biol Chem 253:2769–2776
110 Rinderknecht E, Humbel RE 1978 Primary structure of human insulin-like growth factor II. FEBS Lett 89:283–286.
[111] Isaacs N, James R, Niall H, Bryant-Greenwood G, Dodson G, Evans A, North AC 1978 Relaxin and its structural relationship to insulin. Nature 271:278–281.
[112] Kawakami A, Kataoka H, Oka T, Mizoguchi A, Kimura- Kawakami M, Adachi T, Iwami M, Nagasawa H, Suzuki A, Ishizaki H 1990 Molecular cloning of the *Bombyx mori* prothoracicotropic hormone. Science 247:1333–1335.
[113] Smit AB, Vreugdenhil E, Ebberink RH, Geraerts WP, Klootwijk J, Joosse J 1988 Growthcontrolling molluscan neurons produce the precursor of an insulin-related peptide. Nature 331:535–538.
[114] Le Roith D. Insulin-like growth factors. N. Engl. J. Med. 336 (9), 633-640 (1997).
[115] Baserga, R. Controlling IGF-receptor function: a possible strategy for tumor therapy. *Trends Biotechnol.* 14, 150–152 (1996).
[116] Adams TE, Epa VC, Garrett TP, Ward CW. Cell Mol Life Sci. 2000 Jul;57(7):1050-93.
[117] Brüning JC, et al. Role of brain insulin receptor in control of body weight and reproduction. Science. 2000 Sep 22;289(5487):2122-5.
[118] Zapf J, Froesch ER 1986 Insulin-like growth factors/somatomedins: structure, secretion, biological actions and physiological role. Horm Res 24:121–130.
[119] Daughaday WH, Rotwein P 1989 Insulin-like growth factors I and II. Peptide messenger ribonucleic acid and gene structures, serum, and tissue concentrations. Endocr Rev10:68–91
[120] Shimatsu A, Rotwein P 1987 Mosaic evolution of the insulin-like growth factors. Organization, sequence, and expression of the rat insulin-like growth factor I gene. J Biol Chem 262:7894–7900.
[121] Kajimoto Y, Rotwein P 1991 Structure of the chicken insulin-like growth factor I gene reveals conserved promoter elements. J Biol Chem 266:9724–9731.
[122] . Perfetti R, Scott LA, ShuldinerAR1994 The two nonallelic insulinlike growth factor-I genes in *Xenopus laevis* are differentially regulated during development. Endocrinology 135:2037–2044.
[123] Koval A, Kulik V, Duguay S, Plisetskaya E, Adamo ML, Roberts Jr CT, Leroith D, Kavsan V 1994 Characterization of a salmon insulin-like growth factor I promoter. DNA Cell Biol 13:1057–1062.
[124] Chan SJ, Cao QP, Steiner DF 1990 Evolution of the insulin superfamily: cloning of a hybrid insulin/insulin-like growth factor cDNA from amphioxus. Proc Natl Acad Sci USA 87:9319–9323.
[125] Bichell DP, Kikuchi K, Rotwein P 1992 Growth hormone rapidly activates insulin-like growth factor I gene transcription *in vivo*. Mol Endocrinol 6:1899–1908.
[126] Thissen JP, Ketelslegers JM, Underwood LE 1994 Nutritional regulation of the insulin-like growth factors. Endocr Rev 15:80–101.

[127] D'Ercole AJ 1987 Somatomedins/insulin-like growth factors and fetal growth. J Dev Physiol 9:481–495.
[128] Yakar S, Leroith D, Brodt P. The role of the growth hormone/insulin-like growth factor axis in tumor growth and progression: Lessons from animal models. Cytokine Growth Factor Rev 2005;16:407– 420.
[129] Jones JI, Clemmons DR 1995 Insulin-like growth factors and their binding proteins: biological actions. Endocr Rev 16:3–34.
[130] Firth SM, Baxter RC 2002 Cellular actions of the insulin-like growth factor binding proteins. Endocr Rev 23:824–854.
[131] Ullrich, A. et al. Insulin-like growth factor I receptor primary structure: comparison with insulin receptor suggests structural determinants that define functional specificity. *EMBO J* 5, 2503-2512 (1986).
[132] van der Geer P, Hunter T, Lindberg RA 1994 Receptor proteintyrosine kinases and their signal transduction pathways. Ann Rev Cell Biol 10:251–337.
[133] Guan KL 1994 The mitogen activated protein kinase signal transduction pathway: from the cell surface to the nucleus. Cell Signal 6:581–589.
[134] Chakravarthy MV, Abraha TW, Schwartz RJ, Fiorotto ML, Booth FW 2000 Insulin-like growth factor-I extends in vitro replicative life span of skeletal muscle satellite cells by enhancing G1/S cell cycle progression via the activation of phosphatidylinositol 3_-kinase/ AKT signaling pathway. J Biol Chem 275:35942–35952.
[135] Soos MA, Whittaker J, Lammers R, Ullrich A, Siddle K 1990 Receptors for insulin and insulin-like growth factor-I can form hybrid dimers. Characterisati.
[136] Pandini, G. et al. Insulin/IGF1 hybrid receptors have different biological characteristics depending on the insulin receptor isoform involved. *J. Biol. Chem.* 2002 Jul 22.
[137] Cheatham, B. & Kahn, C. R. Insulin action and the insulin signaling network. *Endocr Rev* 16, 117-142 (1995).
[138] White, M. F. The IRS-signaling system: a network of docking proteins that mediate insulin action. *Mol Cell Biochem* 182, 3-11 (1998).
[139] Schubert, M. et al. Insulin receptor substrate-2 deficiency impairs brain growth and promotes tau phosphorylation. *J Neurosci* 23, 7084-7092 (2003).
[140] Numan, S. & Russell, D. S. Discrete expression of insulin receptor substrate-4 mRNA in adult rat brain. *Brain Res Mol Brain Res* 72, 97-102 (1999).
[141] Virkamäki, A., Ueki, K. & Kahn, C. R. Protein-protein interaction in insulin signaling and the molecular mechanisms of insulin resistance. *J Clin Invest* 103, 931-943 (1999).
[142] Jiang, Z. Y. et al. Insulin signaling through AKT/protein kinase B analyzed by small interfering RNA-mediated gene silencing. *Proc Natl Acad Sci USA* 100, 7569-7574 (2003).
[143] Naïmi, M. et al. Nuclear forkhead box O1 controls and integrates key signaling pathways in hepatocytes. *Endocrinology* 148, 2424-2434 (2007).
[144] Frescas, D., Valena, L. & Accili, D. Nuclear trapping of the forkhead transcription factor FoxO1 via Sirt-dependent deacetylaaon promotes expression of glucogenetic genes. *J Biol Chem* 280, 20589-20595 (2005).
[145] Sarbassov DD, Ali SM, Sabatini DM. (2005). Growing roles for the mTOR pathway. Curr Opin Cell Biol. 2005 Dec;17(6):596-603. Epub 2005 Oct 13.
[146] Kaeberlein, M., Powers, R.W. III, Steffen, K.K., Westman, E.A., Hu, D., Dang, N., Kerr, E.O., Kirkland, K.T., Fields, S., and Kennedy, B.K. (2005). Regulation of yeast replicative life span by TOR and Sch9 in response to nutrients. Science *310*, 1193–1196.
[147] Powers, R.W. III, Kaeberlein, M., Caldwell, S.D., Kennedy, B.K., and Fields, S. (2006). Extension of chronological life span in yeast by decreased TOR pathway signaling. Genes Dev. *20*, 174–184.
[148] White, M. F. Insulin signaling in health and disease. *Science* 302, 1710-1711 (2003).
[149] Kolch, W. Meaningful relationships: the regulation of the Ras/Raf/MEK/ERK pathway by protein interactions. *Biochem J* 351 Pt 2, 289-305 (2000).
[150] Wilcox, G. Insulin and insulin resistance. The Clinical biochemist Reviews/Australian Association of Clinical Biochemists 26, 19-39 (2005).
[151] Sedivy, J., and A. Joyner. *Gene Targeting*. New York: Freeman,1992.

[152] Sternberg, N., and D. Hamilton. Bacteriophage P1 sitespecific recombination. I. Recombination between loxP sites. *J. Mol. Biol.* 150: 467–486, 1981.
[153] Hoess, R., A. Wierzbicki, and K. Abremski. The role of the spacer region in P1 sitespecific recombination. *Nucleic Acids Res.* 14: 2287–2300, 1986.
[154] Abremski, K., and R. Hoess. Bacteriophage P1 site-specific recombination. *J. Biol. Chem.* 259: 1509–1514, 1984.
[155] Guo, F., D. Gopaul, and G. Van Dyne. Structure of Cre recombinase complexed with DNA in a site-specific recombination synapse. *Nature* 389: 40–46, 1997.
[156] Sudhof, T. C., Czernik, A. J., Kao, H., Takei, K., Johnston, P. A., Horiuchi, A., Wagner, M., Kanazir, S. D., Perin, M. S., De Camilli, P. & Greengard, P. (1989) Science 245, 1474-1480.
[157] Valtorta, F., Iezzi, N., Benfenati, F., Lu, B., Poo, M.-m. & Greengard, P. (1995) Eur. J. Neurosci. 7, 261-270.
[158] Lu, B., Greengard, P. & Poo, M.-m. (1992) Neuron 8, 521-529.
[159] Chin, L.-S.; Li, L.; Ferreira, A.; Kosik, K. S.; Greengard, P. : Impairment of axonal development and of synaptogenesis in hippocampal neurons of synapsin I-deficient mice. *Proc. Nat. Acad. Sci.* 92: 9230-9234, 1995.
[160] Zhu Y, Romero MI, Ghosh P, Ye Z, Charnay P, Rushing EJ, Marth JD, Parada LF. (2001). Ablation of NF1 function in neurons induces abnormal development of cerebral cortex and reactive gliosis in the brain. Genes Dev. 2001 Apr 1;15(7):859-76.
[161] Hsiao, K. K. *et al.* Correlative memory deficits, Aelevation and amyloid plaques in transgenic mice. *Science* 274, 99–102 (1996).
[162] Irizarry, M. C., McNamara, M., Fedorchak, K., Hsiao, K. K. & Hyman, B. T. APPSW transgenic mice develop age-related Adeposits and neuropil abnormalities, but no neuronal loss in CA1. *J. Neuropathol. Exp. Neurol.* 56, 965–973 (1997).
[163] Frautschy, S. A. *et al.* Microglial response to amyloid plaques in APPSW transgenic mice. *Am. J. Pathol.* 152, 307–317 (1998).
[164] Carlson GA, Borchelt DR, Dake A, Turner S, Danielson V, Coffin JD, Eckman C, Meiners J, Nilsen SP, Younkin SG, Hsiao KK. (1997). Genetic modification of the phenotypes produced by amyloid precursor protein overexpression in transgenic mice. Hum Mol Genet. 1997 Oct;6(11):1951-9.
[165] Bothe Gerald W. M. Phenotyping of APP-SWE Transgenic Mice: Mortality, Neuroanatomy, Behavior. Neuroscience Conference, Washington, DC, Program # 83.6 (2005).
[166] Meilandt, W. J., Cisse, M., Ho, K., Wu, T., Esposito, L. A., Scearce-Levie, K., Cheng, I. H., Yu, G. Q., and Mucke, L. (2009) Neprilysin Overexpression Inhibits Plaque Formation But Fails to Reduce Pathogenic A{beta} Oligomers and Associated Cognitive Deficits in Human Amyloid Precursor Protein Transgenic Mice. *J. Neurosci.* 29, 1977-1986.
[167] Matsubara, E., Bryant-Thomas, T., Pacheco, Q. J., Henry, T. L., Poeggeler, B., Herbert, D., Cruz-Sanchez, F., Chyan, Y. J., Smith, M. A., Perry, G., Shoji, M., Abe, K., Leone, A., Grundke-Ikbal, I., Wilson, G. L., Ghiso, J., Williams, C., Refolo, L. M., Pappolla, M. A., Chain, D. G., and Neria, E. (2003) Melatonin increases survival and inhibits oxidative and amyloid pathology in a transgenic model of Alzheimer's disease. *J Neurochem.* 85, 1101-1108.
[168] El Khoury, J., Toft, M., Hickman, S. E., Means, T. K., Terada, K., Geula, C., and Luster, A. D. (2007) Ccr2 deficiency impairs microglial accumulation and accelerates progression of Alzheimer-like disease. *Nat. Med.* 13, 432-438.
[169] Hoesche C, Sauerwald A, Veh RW, Krippl B, Kilimann MW. (1993). The 5'-flanking region of the rat synapsin I gene directs neuron-specific and developmentally regulated reporter gene expression in transgenic mice. J Biol Chem. 1993 Dec 15;268(35):26494-502.
[170] Kappeler L, De Magalhaes Filho CM, Dupont J, Leneuve P, Cervera P, Périn L, Loudes C, Blaise A, Klein R, Epelbaum J, Le Bouc Y, Holzenberger M. (2008). Brain IGF-1 receptors control mammalian growth and lifespan through a neuroendocrine mechanism. PLoS Biol. 2008 Oct 28;6(10).
[171] Burks DJ, Font de Mora J, Schubert M, Withers DJ, Myers MG, Towery HH, Altamuro SL, Flint CL, White MF. (2000). IRS-2 pathways integrate female reproduction and energy homeostasis. Nature. 2000 Sep 21;407(6802):377-82.

[172] Finlay BL, Darlington RB (1995) Linked regularities in the development and evolution of mammalian brains. Science 268:1578–1584.
[173] Schubert M, Brazil DP, Burks DJ, Kushner JA, Ye J, Flint CL, Farhang-Fallah J, Dikkes P, Warot XM, Rio C, Corfas G, White MF. (2003). Insulin receptor substrate-2 deficiency impairs brain growth and promotes tau phosphorylation. J Neurosci. 2003 Aug 6;23(18):7084-92.
[174] Liu JP, Baker J, Perkins JA, Robertson EJ, Efstratiadis A (1993) Mice carrying null mutations of the genes encoding insulin-like growth factor I (Igf-1) and type 1 IGF receptor (Igf1r). Cell 75:59–72.
[175] Walsh JS, Welch HG, Larson EB. Survival of outpatients with Alzheimer-type dementia. Ann Intern Med 1990; 113:429–34.
[176] Ott, A., Stolk, R. P., van Harskamp, F., Pols, H. A., Hofman, A., and Breteler, M. M. (1999) Diabetes mellitus and the risk of dementia: The Rotterdam Study. Neurology 53, 1937-1942.
[177] Luchsinger, J. A., Tang, M. X., Shea, S., and Mayeux, R. (2004) Hyperinsulinemia and risk of Alzheimer disease. Neurology 63, 1187-1192.
[178] Haan, M. N. (2006) Therapy Insight: type 2 diabetes mellitus and the risk of lateonset Alzheimer's disease. Nat. Clin. Pract. Neurol. 2, 159-166.
[179] Stewart, R., and Liolitsa, D. (1999) Type 2 diabetes mellitus, cognitive impairment and dementia. Diabet. Med. 16, 93-112.
[180] Frolich, L., Blum-Degen, D., Bernstein, H. G., Engelsberger, S., Humrich, J., Laufer, S., Muschner, D., Thalheimer, A., Turk, A., Hoyer, S., Zochling, R., Boissl, K. W., Jellinger, K., and Riederer, P. (1998) Brain insulin and insulin receptors in aging and sporadic Alzheimer's disease. J. Neural Transm. 105, 423-438.
[181] Frolich, L., Blum-Degen, D., Riederer, P., and Hoyer, S. (1999) A disturbance in the neuronal insulin receptor signal transduction in sporadic Alzheimer's disease. Ann. N. Y. Acad. Sci. 893, 290-293.
[182] Rivera, E. J., Goldin, A., Fulmer, N., Tavares, R., Wands, J. R., and de la Monte, S. M. (2005) Insulin and insulin-like growth factor expression and function deteriorate with progression of Alzheimer's disease: link to brain reductions in acetylcholine. J. Alzheimers Dis. 8, 247-268.
[183] Karen Hsiao, * Paul Chapman, Steven Nilsen, Chris Eckman, Yasuo Harigaya, Steven Younkin, Fusheng Yang, Greg Cole. (1996). Correlative memory deficits, Abeta elevation, and amyloid plaques in transgenic mice. Science 4 October 1996:Vol. 274. no. 5284, pp. 99 - 103.
[184] Jada Lewis,* Dennis W. Dickson,* Wen-Lang Lin, Louise Chisholm, Anthony Corral, Graham Jones, Shu-Hui Yen, Naruhiko Sahara, Lisa Skipper, Debra Yager, Chris Eckman, John Hardy, Mike Hutton, Eileen McGowan. (2001). Enhanced neurofibrillary degeneration in transgenic mice expressing mutant tau and APP. Science 24 August 2001: Vol. 293. no. 5534, pp. 1487 – 1491.
[185] Rempe D, Vangeison G, Hamilton J, Li Y, Jepson M, Federoff HJ. (2006) Synapsin I Cre transgene expression in male mice produces germline recombination in progeny. Genesis. 2006 Jan;44(1):44-9.
[186] Withers DJ, Gutierrez JS, Towery H, Burks DJ, Ren JM, Previs S, Zhang Y, Bernal D, Pons S, Shulman GI, Bonner-Weir S, White MF.(1998) Disruption of IRS-2 causes type 2 diabetes in mice. Nature. 1998 Feb 26;391(6670):900-4.
[187] Selman C, Lingard S, Choudhury AI, Batterham RL, Claret M, Clements M, Ramadani F, Okkenhaug K, Schuster E, Blanc E, Piper MD, Al-Qassab H, Speakman JR, Carmignac D, Robinson IC, Thornton JM, Gems D, Partridge L, Withers DJ. (2008). Evidence for lifespan extension and delayed age-related biomarkers in insulin receptor substrate 1 null mice. FASEB J. 2008 Mar;22(3):807-18. Epub 2007 Oct 10.
[188] Doria A, Patti ME, Kahn CR,(2008). The emerging genetic architecture of type 2 diabetes. Cell Metab. 2008 Sep;8(3):186-200.

[189] Toyama H, Ye D, Ichise M, Liow JS, Cai L, Jacobowitz D, Musachio JL, Hong J, Crescenzo M, Tipre D, Lu Q, Zoghbi S, Vines DC, Seidel J, Katada K, Green MV, Pike VW, Cohen RM, Innis RB (2005). PET imaging of brain with the beta-amyloid probe, [11C]6-OHBTA-1, in a transgenic mouse model of Alzheimer's disease. Eur J Nucl Med Mol Imaging. 2005 May;32(5):593-600. Epub 2005 Mar 25.

[190] Vloeberghs E, Van Dam D, Franck F, Serroyen J, Geert M, Staufenbiel M, De Deyn PP (2008) Altered ingestive behavior, weight changes, and intact olfactory sense in an APP overexpression model. Behav Neurosci. 2008 Jun;122(3):730-2.

[191] Lalonde R, Lewis TL, Strazielle C, Kim H, Fukuchi K. (2003) Transgenic mice expressing the betaAPP695SWE mutation: effects on exploratory activity, anxiety, and motor coordination. Brain Res. 2003 Jul 4;977(1):38-45.

[192] Wang J, Ho L, Qin W, Rocher AB, Seror I, Humala N, Maniar K, Dolios G, Wang R, Hof PR, Pasinetti GM. (2005) Caloric restriction attenuates beta-amyloid neuropathology in a mouse model of Alzheimer's disease., FASEB J. 2005 Apr;19(6):659-61. Epub 2005 Jan 13

[193] Roth G. S., Lane M. A., Ingram D. K., Mattison J. A., Elahi D., Tobin J. D. et al. (2002) Biomarkers of caloric restriction may predict longevity in humans. Science 297: 811.

[194] Mattison J. A., Lane M. A., Roth G. S. and Ingram D. K. (2003) Calorie restriction in rhesus monkeys. Exp. Gerontol. 38: 35–46.

[195] Bodkin N. L., Ortmeyer H. K. and Hansen B. C. (1995) Longterm dietary restriction in older-aged rhesus monkeys: effects on insulin resistance. J. Gerontol. A. Biol. Sci. Med. Sci. 50: B142–B147.

[196] Wolkow, C. A., Kimura, K. D., Lee, M. S., and Ruvkun, G. (2000) Regulation of C. elegans life-span by insulinlike signaling in the nervous system. Science 290, 147-150.

[197] Clancy, DJ; Gems, D; Harshman, LG; Oldham, S; Stocker, H; Hafen, E; Leevers, SJ; Partridge, L.(2001). Extension of life-span by loss of CHICO, a Drosophila insulin receptor substrate protein. Science 2001, 292: 104-106.

[198] Tatar, M; Kopelman, A; Epstein, D; Tu, MP; Yin, CM; Garofalo, RS. (2001). A mutant Drosophila insulin receptor homolog that extends life-span and impairs neuroendocrine function. Science 2001, 292: 107-110.

[199] Hwangbo, DS; Gershman, B; Tu, MP; Palmer, M; Tatar, M. (2004). Drosophila dFOXO controls lifespan and regulates insulin signalling in brain and fat body. Nature 2004, 429: 562-566.

[200] Sell, C; Dumenil, G; Deveaud, C; Miura, M; Coppola, D; DeAngelis, T; Rubin, R; Efstratiadis, A; Baserga, R. (1994). Effect of a null mutation of the insulin-like growth factor I receptor gene on growth and transformation of mouse embryo fibroblasts. Mol. Cell Biol. 1994, 14: 3604-3612.

[201] Accili, D; Drago, J; Lee, EJ; Johnson, MD; Cool, MH; Salvatore, P; Asico, LD; Jose, PA; Taylor, SI; Westphal, H.(1996). Early neonatal death in mice homozygous for a null allele of the insulin receptor gene. Nat. Genet. 1996, 12: 106-109.

[202] Holzenberger, M; Dupont, J; Ducos, B; Leneuve, P; Geloen, A; Even, PC; Cervera, P; Le Bouc, Y. (2003). IGF-1 receptor regulates lifespan and resistance to oxidative stress in mice. Nature 2003, 421: 182-187.

[203] Selman, C; Lingard, S; Choudhury, AI; Batterham, RL; Claret, M; Clements, M; Ramadani, F; Okkenhaug, K; Schuster, E; Blanc, E; Piper, MD; Al Qassab, H; Speakman, JR; Carmignac, D; Robinson, IC; Thornton, JM; Gems, D; Partridge, L; Withers, DJ. (2008). Evidence for lifespan extension and delayed age-related biomarkers in insulin receptor substrate 1 null mice. FASEB J 2008, 22: 807-818.

[204] Taguchi, A., Wartschow, L. M., and White, M. F. (2007) Brain IRS2 signaling coordinates life span and nutrient homeostasis. Science 317, 369-372.

[205] Lammich, S., Kojro, E., Postina, R., Gilbert, S., Pfeiffer, R., Jasionowski, M., Haass, C., and Fahrenholz, F. (1999) Constitutive and regulated alphasecretase cleavage of Alzheimer's amyloid precursor protein by a disintegrin metalloprotease. Proc Natl Acad Sci U S A 96, 3922-3927.

[206] Leissring, M. A., Farris, W., Chang, A. Y., Walsh, D. M., Wu, X., Sun, X., Frosch, M. P., and Selkoe, D. J. (2003) Enhanced proteolysis of beta-amyloid in APP transgenic mice prevents plaque formation, secondary pathology, and premature death. *Neuron* 40, 1087-1093.
[207] Zhao, L., Teter, B., Morihara, T., Lim, G. P., Ambegaokar, S. S., Ubeda, O. J., Frautschy, S. A., and Cole, G. M. (2004) Insulin-degrading enzyme as a downstream target of insulin receptor signaling cascade: implications for Alzheimer's disease intervention. *J. Neurosci.* 24, 11120-1112.
[208] El Khoury, J., Toft, M., Hickman, S. E., Means, T. K., Terada, K., Geula, C., and Luster, A. D. (2007) Ccr2 deficiency impairs microglial accumulation and accelerates progression of Alzheimer-like disease. *Nat. Med.* 13, 432-438.
[209] Costantini, C., Scrable, H., and Puglielli, L. (2006) An aging pathway controls the TrkA to p75NTR receptor switch and amyloid beta-peptide generation. *EMBO J.* 25, 1997-2006.
[210] Puglielli, L. (2008) Aging of the brain, neurotrophin signaling, and Alzheimer's disease: is IGF1-R the common culprit? *Neurobiol. Aging* 29, 795-811.
[211] Puglielli, L., Ellis, B. C., Saunders, A. J., and Kovacs, D. M. (2003) Ceramide stabilizes beta-site amyloid precursor protein-cleaving enzyme 1 and promotes amyloid beta-peptide biogenesis. *J. Biol. Chem.* 278, 19777-19783.
[212] Meilandt, W. J., Cisse, M., Ho, K., Wu, T., Esposito, L. A., Scearce-Levie, K., Cheng, I. H., Yu, G. Q., and Mucke, L. (2009) Neprilysin Overexpression Inhibits Plaque Formation But Fails to Reduce Pathogenic A{beta} Oligomers and Associated Cognitive Deficits in Human Amyloid Precursor Protein Transgenic Mice. *J. Neurosci.* 29, 1977-1986.
[213] Matsubara, E., Bryant-Thomas, T., Pacheco, Q. J., Henry, T. L., Poeggeler, B., Herbert, D., Cruz-Sanchez, F., Chyan, Y. J., Smith, M. A., Perry, G., Shoji, M., Abe, K., Leone, A., Grundke-Ikbal, I., Wilson, G. L., Ghiso, J., Williams, C., Refolo, L. M., Pappolla, M. A., Chain, D. G., and Neria, E. (2003) Melatonin increases survival and inhibits oxidative and amyloid pathology in a transgenic model of Alzheimer's disease. *J Neurochem.* 85, 1101-1108.
[214] Nathan, C., Calingasan, N., Nezezon, J., Ding, A., Lucia, M. S., La Perle, K., Fuortes, M., Lin, M., Ehrt, S., Kwon, N. S., Chen, J., Vodovotz, Y., Kipiani, K., and Beal, M. F. (2005) Protection from Alzheimer's-like disease in the mouse by genetic ablation of inducible nitric oxide synthase. *J Exp. Med* 202, 1163-1169.
[215] Leissring, M. A., Farris, W., Chang, A. Y., Walsh, D. M., Wu, X., Sun, X., Frosch, M. P., and Selkoe, D. J. (2003) Enhanced proteolysis of beta-amyloid in APP transgenic mice prevents plaque formation, secondary pathology, and premature death. *Neuron* 40, 1087-1093.
[216] Cohen, E., Bieschke, J., Perciavalle, R. M., Kelly, J. W., and Dillin, A. (2006)Opposing activities protect against age-onset proteotoxicity. *Science* 313, 1604-1610.
[217] Araki E, Lipes MA, Patti ME, Brüning JC, Haag B 3rd, Johnson RS, Kahn CR. (1994) Alternative pathway of insulin signalling in mice with targeted disruption of the IRS-1 gene. Nature. 1994 Nov 10;372(6502):128-9.
[218] Withers DJ, Gutierrez JS, Towery H, Burks DJ, Ren JM, Previs S, Zhang Y, Bernal D, Pons S, Shulman GI, Bonner-Weir S, White MF.(1998), Disruption of IRS-2 causes type 2 diabetes in mice. Nature. 1998 Feb 26;391(6670):900-4.
[219] Kido Y, Burks DJ, Withers D, Bruning JC, Kahn CR, White MF, Accili D. (2000). Tissuespecific insulin resistance in mice with mutations in the insulin receptor, IRS-1, and IRS-2. J Clin Invest. 2000 Jan;105(2):199-205.
[220] Aspinwall CA, Qian WJ, Roper M, Kulkarni RN, Kahn CR, Kennedy RT: Roles of insulin receptor substrate-1, phosphatidylinositol 3-kinase, and release of intracellular Ca2+ stores in insulin-stimulated insulin secretion in _-cells. *J Biol Chem* 275:22331–22338, 2000.
[221] Xu GG, Gao ZY, Borge PDJ, Wolf BA: Insulin receptor substrate 1-induced inhibition of endoplasmic reticulum Ca2+ uptake in _-cells. *J Biol Chem* 274: 12067–12074, 1999.
[222] Brüning JC, Winnay J, Cheatham B, Kahn CR.(1997). Differential signaling by insulin receptor substrate 1 (IRS-1) and IRS-2 in IRS-1-deficient cells. Mol Cell Biol. 1997 Mar;17(3):1513-21.

[223] Feyt C, Kienlen-Campard P, Leroy K, N'Kuli F, Courtoy PJ, Brion JP, Octave JN. (2005). Lithium chloride increases the production of amyloid-beta peptide independently from its inhibition of glycogen synthase kinase 3. J Biol Chem. 2005 Sep 30;280(39):33220-7.
[224] Phiel CJ, Wilson CA, Lee VM, Klein PS. (2003). GSK-3alpha regulates production of Alzheimer's disease amyloid-beta peptides. Nature. 2003 May 22;423(6938):435-9.

Acknowledgments

My gratitude goes to Dr. Markus Schubert for providing me with this project in his lab.

Also I would like to acknowledge and thank Dr. Michael Udelhoven for his support and supervision the last years, my colleagues for their help in the lab and particularly Dr. Katharina Schillbach and Uschi Leeser for joining and helping me in the mice facility the last few years.
Oliver Stöhr I would like to thank for being all the years a good friend.

Furthermore I thank my father for his great support and advice.

Finally I would like to thank my "Mäusgen" Simone for her love, patience and assistance.

Die VDM Verlagsservicegesellschaft sucht für wissenschaftliche Verlage abgeschlossene und herausragende

Dissertationen, Habilitationen, Diplomarbeiten, Master Theses, Magisterarbeiten usw.

für die kostenlose Publikation als Fachbuch.

Sie verfügen über eine Arbeit, die hohen inhaltlichen und formalen Ansprüchen genügt, und haben Interesse an einer honorarvergüteten Publikation?

Dann senden Sie bitte erste Informationen über sich und Ihre Arbeit per Email an *info@vdm-vsg.de*.

Sie erhalten kurzfristig unser Feedback!

VDM Verlagsservicegesellschaft mbH
Dudweiler Landstr. 99   Telefon +49 681 3720 174
D - 66123 Saarbrücken   Fax   +49 681 3720 1749

www.vdm-vsg.de

Die VDM Verlagsservicegesellschaft mbH vertritt

Printed by Books on Demand GmbH, Norderstedt / Germany